Preface

Smart places, such as the dematerialization of diverse natural ecosystems, involve several autonomous ecosystems that interconnect and promote the integration of information and the convergence of necessarily secure functions and activities that depend on reliable data and reliable sources. The problem of data management, its quality, and its governance is aggravated by the amount of data generated, the multiplicity of devices, spaces, infrastructures, users and connected entities, being a technological and management challenge. The various cyber risks can lead to data compromise, exploitation of weaknesses, infiltration of systems, conditioning the functioning of the city and, in the limit, disengaging or even destroying the physical infrastructure to the point where citizens have their lives threatened.

The research methodology chosen for this work is the Design Science Research (DSR) methodology, in the problem-centered approach, where we intend to construct an artifact, which allows us to evaluate viable alternatives for using reliable, blockchain-based technology.

The proposal focuses on a generic data model to be applied to smart places in the context of smart cities, focusing on their revision and structuring in data management aspects and their governance. The proposed model adopts the blockchain technologies and applies to the different characteristics of the city, in the electronic governance, in the contracting of products and services and the collection of data. Various IoT objects and multiple networks, along with blockchain technology, can result in safer and more efficient spaces and cities. This work explores the concept of smart cities in the mobility and transport ecosystem, using blockchain technology as a platform for data security and reliability, applied in the ticketing subsystem and traffic subsystem, for the safety and control of the logs generated by the numerous devices.

With this artifact it is intended the generalization of the model to be applied to different subsystems allowed that generic data models, be integrated and automated, with quality data and reliable information. Controlling data flows, managing the data and information lifecycle will enable a more reliable data management, information management, and governance process. I dedicate this work to my wife and children, to my father and mother.

The first thanks go to my supervisor, Prof. Henrique São Mamede, and to my co-supervisor, Prof. Ramiro Gonçalves, for the objective, planned and challenging path that led me to this path. To the Universities, Universidade Aberta and Universidade de Trás-os-Montes e Alto Douro, for the demanding and excellent Ph.D. Program of Web Science and Technology, multidisciplinary that allowed an enrichment far beyond the course itself.

My thanks to my friends, my co-workers, and my company. To my colleagues who started this journey and set out on its first edition.

General index

PREFACE I
GENERAL INDEX III
TABLE INDEX V
FIGURE INDEX VII
LIST OF ABBREVIATIONS, ACRONYMS AND ACRONYMS VIII
I.   INTRODUCTION 1
    1.   RESEARCH PROBLEM 7
    2.   GOALS 7
        2.1.   MAIN GOAL 7
        2.2.   SPECIFIC OBJECTIVES 7
    3.   JUSTIFICATION 8
    4.   STRUCTURE OF THE THESIS 8
II.   THEORETICAL FRAMEWORK 12
    1.   *SMART PLACES* 12
        1.1.   *SMART CITY* 13
        1.2.   SMART CITY MODELS 19
        1.3.   *SMART CITIES* AND IOT (INTERNET OF THINGS) 31
        1.4.   *BIG DATA* 36
    2.   BLOCKCHAIN 39
        2.1.   CHARACTERISTICS 42
        2.2.   MERKLE TREE 49
        2.3.   BLOCKCHAIN PLATFORMS 51
        2.4.   BLOCKCHAIN COMPARISON WITH OTHER TECHNOLOGIES IN IOT ENVIRONMENT ... 57
        2.5.   ARCHITECTURES AND TAXONOMY 60
        2.6.   BLOCKCHAIN TECHNOLOGY APPLICATIONS 64
    3.   DATA MARKET 66
    4.   MOBILITY ECOSYSTEM 73
        4.1.   SYSTEMS IN PASSENGER RAIL TRANSPORT 74
        4.2.   INFORMATION TECHNOLOGIES IN PUBLIC TRANSPORT SYSTEMS ... 79
        4.3.   TICKETING SYSTEM 91
III. EMPIRICAL STUDY 96
    1.   METHODOLOGY 96
        1.1.   METHOD 101
        1.2.   PHASES 102

	1.3.	CENTRAL PROBLEM 105
2.		SEARCH RESULTS 105
	2.1.	CHOOSING BLOCKCHAIN TECHNOLOGY 106
	2.2.	GENERIC DATA MODEL 114
	2.3.	GENERAL ARCHITECTURE 117
	2.4.	APPLICATION OF THE METHODOLOGY 119
	2.5.	THE CHOICE OF THE MOBILITY ECOSYSTEM AND THE APPLICATION OF TICKETING 120
	2.6.	ARTIFACT DESIGN AND DEVELOPMENT 121
	2.7.	ASSESSMENT (5TH STEP DSRM METHODOLOGY) 136
	2.8.	COMMUNICATION (6TH DSRM STEP) 137
	2.9.	PRACTICAL RESULTS 138
3.		DISCUSSION OF RESULTS 140
	3.1.	RELATED WORKS 145
IV.		FINAL CONSIDERATIONS 148
1.		CONCLUSIONS 150
2.		LIMITATIONS AND FUTURE WORK 152
BIBLIOGRAPHY 153
ANNEX I - COMPARATIVE TABLE OF BLOCKCHAIN PLATFORMS I

Table Index

Table II.1 - Comparison of two smart cities metrics models 20

Table II.2 - Smart city technology rating standard 22

Table II.3 - "Smart" city solutions to move from data to services. 23

Table II.4 - Smart city API solutions 23

Table II.5 - Modeling approaches for smart cities 25

Table II.6 - Smart city benchmarking tools 26

Table II.7 - Security standards and norms and recommendations for smart places cybersecurity 27

Table II.8 - Smart city indicators 29

Table II.9 - Comparison of standard protocols used in IoT and industry 4.0 36

Table II.10 - Characteristics of consensus algorithms 44

Table II.11 - Blockchain information security analysis 52

Table II.12 - Analysis of blockchain performance characteristics 53

Table II.13 - IoT layered analytics 57

Table II.14 - Current challenges in IoT security and mitigation measures 58

Table II.15 - Blockchain project related design decisions 62

Table II.16 - Blockchain project related to design decisions. 63

Table II.17 - Blockchain projects related to project decisions about blockchain configuration. 64

Table II.18- Light rail acquisition trends 75

Table II.19 - Vulnerability Factors in URT 78

Table II.20 - Summary of the effectiveness of strategies 82

Table II.21 - Degrees of Automation 87

Table II.22 - Main functions of CBTC 88

Table III.1 - Contributions to the DSR methodology 96

Table III.2– Blockchain vs. Centralized legacy database and distributed database 107

Table III.3 – Compare nodes between permissionless, permissioned and centralized database blockchains. 107

Table III.4 - Strengths and weaknesses of blockchain technology 109

Table III.5 - Comparison between PKI and stochastic blockchain based data schema with verification scheme 111

Table III.6 - Drivers for the adoption of Blockchain technology 112

Table III.7 - Blockchain Application Types (BC) 116

Table III.8 - Application of blockchain in the context of smart cities 117

Table III.9 - Fields exported to the Log file 127

Table III.10 - Type of Events considered..127

Table III.11 – Example of the log file..129

Figure Index

Figure II.1 - Overview of the Data-driven Ecosystem and its components 2

Figure II.2 - Merkel's tree 2

Figure II.3 - Design process for blockchain supported systems 2

Figure II.4 - Blockchain based data market 2

Figure II.5 - General data marketplace model 2

Figure II.6 - Simplified data marketplace model 2

Figure II.7 - Data market model and actors 2

Figure II.8 - Information System to Support Planning and Management 2

Figure II.9 - Exploration Support Information System 2

Figure II.10 - Generic Traffic Management Support Information System 2

Figure II.11 - Rolling Stock Support Information System (MC) 2

Figure II.12 - General Architecture of the Ticketing System 2

Figure III.1 - DSR cycles 2

Figure III.2 Context model of a smart city 2

Figure III.3 - Data flows in a smart place 2

Figure III.4 – Macro Flows Intra and Inter smart places, citizen/user centered 2

Figure III.5 - Data flow of alarming and monitoring events 2

Figure III.6 - Artifact object data flow phase 2

Figure III.7 - Artifact Design in the 1st Iteration 2

Figure III.8 - Log control blockchain 2

Figure III.9 - Phase of the Insertion Data Flow in the Database 2

Figure III.10 - Insertion with per-register hash generation 2

Figure III.11 - Artifact demonstration scenario (2nd iteration) 2

Figure III.12 - Simplified Data Structure of Blocks in the 2nd Iteration 2

Figure III.13 - Data structure with Merkel tree 2

Figure III.14 - Main Data Groups/Functions 2

Figure III.15 - Data Flow considering an IoT network and a Blockchain network 2

Figure III.16 - Data Flow Integrating IoT with Blockchain 2

List of Abbreviations, Acronyms and Acronyms

ADR - Action Design Research
AES - Advanced Encryption Standard
AHP - Analytical Hierarchy Process
AI - Artificial Intelligence
AMQP - Advanced Message Queuing Protocol
API's - Application Program Interfaces
ASICs - Application Specific Integrated Circuits
ATC - Automatic Train Control
ATO - Automatic Train Operation
ATP - Automatic Train Protection
ATR - Automatic Train Regulation
ATS - Automatic Train Supervision
BCS - Blockchain Structures
BFT - Byzantine Fault-Tolerant
BI - Business Intelligence
BIM - Building Information Model
BL - Blockchain Local
BMS - Building Management Systems
BREEAM - Building Research Establishment Environmental Assessment Method
BTM - Balise transmission module
C-ITS - Cooperative Intelligent Transport Systems
CBIS - Compute-based information systems
CBTC - Communications-Based Train Control
CCM - Cipher Block Chaining-Message Authentication Code
CDIM - City District Information Model
CI - Computer Interlocking
CIM - City Information Modeling
CityGML - City Geography Markup Language
CLIs - Command-Line Interfaces
CoAP - Constrained Application Protocol
CoT - Cloud of Thing
CSSP - Cloud Storage Provider
CVIM - Common Vehicle Information Model
DAG - Directed Acyclic Graph
DAOs - Decentralized Autonomous Organization
DDS - Data Distribution Service
DHT - Distributed Hash-Table
DLT - Distributed Ledger Technology
dPoS - Delegated Proof of Stake,
DR - Demand Response
DRL - Deep Reinforcement Learning
DRU - Data Recording Unit
DS - Design Science
DSM - Demand Side Management
DSR - Design Science Research
DSRIS - Design Science Research in Information Systems
DSRM - Design Science Research Methodology
DSRPM - DSR Process Model

DTLS - Datagram Transport Layer Security
E2E - End-to-End
ECC - Elliptic Curve Cryptosystems
ECDH - Elliptic-curve Diffie-Hellman
ECDSA - Elliptic Curve Digital Signature Algorithm
ECMV - Elliptic Curve Menezes Vanstone
ERP - Enterprise Resource Planning
ESP - Encapsulation Security Payload
ETCS - European Train Control System
EVM - Ethereum Virtual Machine
FI - Future of Internet
GHOST - Greedy Heaviest-Observed Sub-Tree
GIS - Geographic Information Systems
GoA - Grades of Automation
GPG - GNU Privacy Guard
GPS - Global Positioning System
HLF - Hyperledger Fabric
HMAC - hash message authentication code
IBIS - Integrated On-board Information System
ICD - Implantable Cardioverter Defibrillator
IFC - Industry Foundation Classes
IMDs - Implantable Medical Devices
IoT - Internet of Things
IoV - Internet of Vehicles
IPFS - Inter Planetary File System
IPLD - InterPlanetary Linked Data
IPNS - InterPlanetary Naming System
IPRS - InterPlanetary Record System
IPU - Interlocking Processing Unit
ISD - Information Systems Development
ISDT - Information Systems Design Theory
ISM - Interpretative Structural Modeling
ITS - Intelligent Transport Systems
KMS - Key Management System
KPIs - Key Performance Indicator
KSI - Keyless Signature Infrastructure
LRV - Light Rail Vehicles
M2M - Machine to Machine
MC - Rolling Stock
MDM - Master Data Management
Merkle - Merkle tree root hash
ML - Machine Learning
MPC - Multi-Party Computation
MQTT - Message Queue Telemetry Transport
MQTT-SN - MQTT for Sensor networks
MV - Virtual Machines
NGTC - Next Generation Train Control
NIS - Network and Information Security Directive
NN - Nearest Neighbor
OBC - On-Board Computer

OCS - Object Controller System
OPC UA - Open Platform Communications Unified Architecture
PADR - Participatory Action Design Research
PBFT - Practical Byzantine Fault Tolerance
PDD - Protocol Driven Development
PIS - Passenger Information System
PLS - Physical Layer Security
PN - Level Crossings
PoS - Proof of Stake
PoW - Proof of Work
PKI - Public Key Infrastructure
PUF - Physical Unclonable Functions
QFD - Quality Function Deployment
QoS - Quality of Service
QR - Quick Response
RDF - Resource Description Framework
RE-CAWAR - Requirement Engineering Context Awareness
REST - Representational State Transfer
RFID - Radio-Frequency IDentification
Ripple - Ripple Protocol Consensus Algorithm
ROS - Robot Operating System
SCADA - Supervisory Control and Data Acquisition
SCALE - Smart city Application Ecosystem
SCP - Stellar Consensus Protocol
SDE - Secure Data Exchange
SDN - Software Defined Network
SDRM - Systems Development Research Methodology
SDS - Soft Design Science
SDSM - Soft Design Science Methodology
IS - Information Systems
SIEM - Security Information and Event Management
GIS - Signaling System
SIP - Public Information System
SLA - Service Level Agreement
SOA - Service Oriented Architecture
SPV - Simplified Payment Verification
ICT - Information and Communication Technologies
TLS - Transport Layer Security
TMS - Traffic Management System
UCC - Urban Carrying Capacity
UIC - International Union of Railways
URT - Urban Rail Transit
URTs - Terminal Remote Unit
UTO - Unattended Train Operation
VA - Added Value
VANETs - Vehicular Adhoc Networks
VM - Virtual Machine
VTCU - Vehicle Train Control Unit
WSNs - Wireless Sensor Network
XML - Extensible Markup Language

XMPP - Extensible Messaging and Presence Protocol
ZC - Zone Controller

I. INTRODUCTION

I. INTRODUCTION

This research work is oriented towards presenting a model of smart places based on secure technologies that guarantee the reliability of the generated data, which support its operation, the generation of information, knowledge, and decision-making.

The model will be evaluated using proofs of concept through artifacts that will allow the validation of the model in the alarmistic component and in the existing monitoring systems in a smart city. The chosen system was the ticketing system belonging to the mobility and transport ecosystem.

This modeling seeks to answer the numerous questions that run through the various domains involved in a smart place, where applications are integrated to ensure trust in data and information.

The smart place of this study will be based on smart city concepts and models.

The technology in analysis and evaluation should be supported by blockchain technology as a way of guaranteeing and controlling data flows and, in this way, the trust of data, information, and transactions.

In this perspective, we seek to review several concepts that will be detailed throughout this work, which include defining the concepts of smart place and smart city, smart city models, blockchain characteristics, applications of blockchain technology, challenges and limitations, and the information system to support urban rail transport systems, deepening the traffic management system and the ticketing system.

The scientific work will be consolidated through the development of artifacts for the application of this model in a specific aspect of the functioning of the smart city, in this case in the mobility and transport system and in the management of ticketing events of the light rail system, through the application of the methodology DSR (Design Science Research), based on problem-centered approach (Peffers, Tuunanen, Rothenberger, & Chatterjee, 2007).

The smart city, in the concept presented here, is not only a physical city but also a digital and virtual one, which has several dimensions, such as its governance, institutions, companies, ecological, social, and economic sustainability policies, mobility, and transport, civic participation of different economic and social agents, innovation policies, ICTs (Information and Communication Technologies) and IS

(Information Systems) that converge towards common strategies (Brandão et al., 2018a).

These strategies must comply with integration, relationship, interaction, participation, access to information frameworks, which imply robust connectivity infrastructures, fiber optic networks, wireless networks, and network integration, which will be able to support the growing needs of the various services and applications, monitoring, control, sensing and IoT devices, with the multiplication of sensors, application actuators, with more adaptive architectures, databases that grow vertiginously, with the treatment of huge amounts of data, through concepts of big data, data mining and information visualization (InfoVis) in order to extract relevant information that reorients strategies, policies, the creation of ecosystems that enhance business and new business models.

In summary, this thesis explores the following areas: smart places in a Web environment, focusing on smart cities as an application context, with the basic theme oriented towards data management, mainly governance and data security, and with the possibility of using blockchain support technology as a reliable technology to guarantee the integrity of information and non-compromise of data.

The smart city is understood in two essential dimensions, in a more material dimension, associated with the physical and natural aspects of a city and its optimization, and in an intangible dimension, associated with aspects of people's well-being, mobility, education, health, innovation, social inclusion, and its governance. This city, this urban space, is vulnerable to data compromise (Popescul & Radu, 2016) and false data injection (K. Zhang et al., 2017). The growing volume of data, the large number of devices, spaces, infrastructures, and connected users, extend the risks and can cause systems to be compromised, with the use of weaknesses that can be transmitted or exploited between systems. This central problem can, in extreme situations, allow the exploitation of weaknesses or the possibility of infection that could compromise the functioning of the city itself and disconnect or even destroy the physical infrastructure to the point that citizens have their lives threatened. (Popescul & Radu, 2016)

Data management and governance of the growing amount of data is a technical and management challenge, particularly as data and the resulting information develop into strategic resources with characteristics that make it difficult to govern. Controlling data flows, managing the life cycle of data and information has become a critical point in the process of data management and information management and its governance.

In a first brief introduction, it refers to the use of blockchain applications in several domains, described in some articles mentioned below:

- The article of Yli-Huumo, Ko, Choi, Park, & Smolander (2016) presents as examples the prototypes of applications developed and suggested for the use of blockchain in other environments, such as the IoT, smart contracts, the "smart" property, the distribution of digital content, the botnet, and the P2P transmission protocols, used in an environment decentralized.
- The article of Dorri, Kanhere, Jurdak, & Gauravaram (2017) describes the main components of a smart home, with IoT devices, local storage, miner, and BL (Local Blockchain) and discusses the various transactions and procedures associated with it, with an analysis of security and privacy. The smart home simulation demonstrates that the overall costs incurred by the described method are low and manageable for low-resource IoT devices and are acceptable given the security and privacy benefits offered.

This brief description of blockchain technology application guides you through some of the critical aspects of IoT devices, such as data security and privacy concerns.

Smart places concentrate several aspects common to Organizations, such as spaces for collaboration and competitive "intelligence", with dynamic capabilities and information processes, with control frameworks, metrics, and evaluations. Smart cities are necessarily dynamic organizations, with common models and indicators that must lead to the policy formulation process, with reliable data that support information and business processes.

Competitive "intelligence" and the development of dynamic capabilities in Organizations aim to verify whether the stages of the competitive intelligence cycle can constitute stimulating elements for the dynamic capabilities of Organizations (Garcia, 2017). Competitive "intelligence" promotes the perception of change and provides the "intelligence" necessary for the acquisition of knowledge that will be the basis of action, contributing to the continuous reinvention of the business (detect, apprehend and transform). "Intelligence" should not be presented only to facilitate and improve understanding as a single and immutable capacity, but in multiple and modifiable ways. (Schelini, 2006)

Integral sustainable development presents a practical framework (Brown, 2005) that integrates the conceptual and operational landscape of sustainable development and allows identifying the full range of needs and capabilities of individuals and groups and tailoring the specific development response that fits each unique situation. The framework maps and integrates human consciousness and behavior, culture, systems, and the physical environment into a comprehensive and accurate approach to meeting our social, environmental, and economic challenges.

The conceptualization of a smart city has several dimensions, technology, people, and institutions. (Nam & Pardo, 2011) The strategic principles must be aligned with the three main dimensions that pass through the integration of technological mediation infrastructures and services, social learning to strengthen the human base, and governance for institutional improvement and citizen involvement.

A "smart" model (Lazaroiu & Roscia, 2012) can be the support to calculate the indexes and metrics to evaluate a smart city. These indicators, as they are not homogeneous, have to be considered for the policy formulation process and as a point of discussion between stakeholders, as well as citizens, for a final decision on the measures to be adopted and the evaluation of the best options.

The reference models of a smart city can be supported by the innovation characteristics of "smart" ecosystems, with the structuring of smart city notions in ecosystems oriented to green criteria, interconnected, instrumented, open, integrated and innovative, which can compose a planning framework. (Zygiaris, 2013)

Santos (2015) refers to the cities of the future associated with talent, innovation, and collaboration, with the view that the smart city enhances the generation of Added Value.

The city is presented as a customer, "city as a customer" (Amarnath, 2011) where urbanization can create numerous sectoral implications and opportunities. Megatrends such as climate change, demographic change, globalization, and urbanization are global transformative forces that define the future world, impacting businesses, societies, economies, cultures, and lives, and with implications for energy (networks " smart", renewable energies, and energy as a service), in health, in industry, in infrastructure and in cities (transport, mobility and logistics systems, and eco-construction technologies).

The need for an information framework (Jin et al., 2014) can cover the complete urban information system, the level of sensing and networking, the data management support structure, and cloud access, with the integration of systems and services.

The exploration of data, supported by platforms, is in itself a challenge due to the diverse, heterogeneous, and interconnected set of data, to build a unified view of data. (Sadoshi, 2017) The data model presents difficulties at the level of form and metadata from each source, to offer continuous consolidation of data under uncertainty to make the data model inherently adaptable. The value that can be attributed to data can come from different services and forms, from processing and analyzing raw data, to presenting contexts that can trigger the construction of new services and businesses. (Brandão et al., 2019)

In this domain, the difficulty in obtaining reliable data can affect the management of information, compromise sustainable mobility, and can prove to be a problem in modular platforms. These platforms aim to integrate data to manage information, for example, traffic, and to achieve "smart" mobility (Teixeira et al., 2017) although, with dynamic data collection, it is still limited to assess how it would be possible to read and predict levels of traffic congestion and emissions in real-time.

The data-centric perspective describes data management techniques to ensure consistency, granularity, interoperability, and data reuse. This perspective focused on the data lifecycle of a smart city is interdependent on data management in the data security and privacy layers, and on the supporting infrastructure. (Gharaibeh et al., 2017)

Research on software platforms reveals that to facilitate the development, integration, and deployment of applications in smart cities, the most cited enabling technologies are the Internet of Things (IoT), cloud computing, big data, and cyber-physical systems. (Santana et al., 2016)

Cities that intend to be smart cities are based on monitoring and ubiquitous computing and guide the economy and governance towards innovation, creativity, and entrepreneurship, supported by skilled people. These cities, instrumented with digital devices and infrastructure that produce big data, allow real-time analysis of city life, new forms of urban governance, and the basis for idealizing more sustainable, competitive, open, transparent, efficient, and productive cities. (Kitchin, 2014)

Studies on various applications in smart cities are numerous and reveal the high interest and importance of this type of smart place and the impact on sustainable, participatory, and inclusive development, which guide projects and developments in this field.

In this introductory text, some examples of this orientation are revealed below.

- The case study of the SmartSantander experimental platform intended to produce the following objectives: the architectural reference model for IoT experimentation; an experimental, large, scalable, heterogeneous, and reliable installation; a representative set of use cases implemented for the experimental installation; and a large set of future experiments for the Internet (Sanchez et al., 2011). The key aspect was the availability of a large set of applications based on their high potential and impact on citizens, in order to attract the greatest interest and allow to demonstrate the usefulness of the experimental platform.
- A smart city based on Cloud of Things (CoT) can offer a common and global approach to cloud sensors and innovative services to perform the aggregation of heterogeneous resources, defined in the Cloud of Things paradigm. (Petrolo et al., 2012)
- The assisted or supportive smart cities model, through the application of ubiquitous computing in accessibility, provides solutions to support people with disabilities. (Telles, 2017)

This work, which is intended to be developed in the context of smart cities, in the theoretical framework, studies the models, the incorporation of IoT in the city, the use of big data as a tool for analysis and discovery of patterns, provides details on the use of technology blockchain in various domains, its characteristics, block structure, digital signature, consensus protocols/algorithms, verification optimization process through Merkel trees, analyzes blockchain platforms, perspective and contextualizes data markets and presents in the ecosystem mobility, the information systems to support urban rail transport and, in greater detail, the ticketing system.

The empirical study describes the methodology, presents the research results, with the generic data model, the architecture, the application of the methodology, the conception and development of the artifacts, in two iterations that complement each other, the evaluation of the artifacts and the communication of the research results. In the discussion of the results, the data flow object of the research is analyzed and it reflects on the paths that would provide the global control of data flows and other blockchain applications in a smart city. At this point, some works related to the work that was developed are also reviewed.

The final considerations summarize the main conclusions of the work and present the set of limitations that still exist and that may lead to further investigations, the deepening of some aspects of the generic data model, and the control of other phases of the data flow.

1. Research Problem

With this research, we intend to clarify important aspects of the design, development, and management of data in smart cities, mainly on trust in data, in order to answer the following question:

Will it be possible to develop a data and information management platform, in the context of a smart city, based on reliable technology, which avoids compromising data and its origin?

2. Goals

The definition of objectives reveals the need to focus the theme and focus the work on the aspects considered essential so that it allows, on the one hand, to demonstrate through the scientific method based on the most appropriate methodology the proposed answer to the problem and, on the other hand, to try to generalize the solution found.

2.1. Main goal

The general objective of the project is to apply a generic data model, to be proposed, in support of the smart city concept, in order to systematize its actions and the control of data flows and data quality, which allow data management and information, reliably and securely.

2.2. Specific objectives

In summary and below, the following three specific objectives are defined that are intended to be achieved with the completion of the work to be developed.

1. To structure a generic data model to support the smart city concept that leads and allows the alignment of the application of data ecosystems with natural ecosystems.
2. Structuring the relationships between ecosystems, participants, and data that facilitate the use of blockchain technology in data management.
3. Ensure mechanisms of reliability in the management of data and data sources.

3. Justification

The future of the world will be data driven, staying connected, monitored and auditable, listening, seeing and learning.

The internet statistics in 2019[1] indicate around 4.38 billion Internet users as of March 31, 2019, corresponding to 56.8% of the population, with a growth of 1,114% between 2000 and 2019, with 50.1% concentrated in Asia.

In the supply and order of data storage, it is estimated the supply of 24,800 exabytes for orders of 42,700 exabytes in 2020[2].

In the NetworkWorld article[3], the IDC[4] estimates, by 2025, around 175 zettabytes (ZB) of data, of which 90 ZB of data will be created on IoT devices and 49% of the data will be stored in public cloud environments and 30% of the data generated will be consumed in real time. IoT equipment installed in smart cities in 2018[5] were 463.5 million (22% more than in 2017).

Given the importance of data in smart places, the proposed project aims to focus research on data reliability and on the proposal of a generic data model in which its evaluation will be through the development of artifacts, based on proofs of concept, to respond to the control issues of data flows, data and information management, and their governance, proposing the use of blockchain technology to ensure the effectiveness of the model and confidence in data and data flows.

---

[1] Internet Statistics - https://www.internetworldstats.com/stats.htm, accessed on 21-06-2019
[2] statist - https://www.statista.com/statistics/751749/worldwide-data-storage-capacity-and-demand/, accessed on 21-06-2019
[3] Networkworld - https://www.networkworld.com/article/3325397/idc-expect-175-zettabytes-of-data-worldwide-by-2025.html, accessed on 21-06-2019
[4] Seagate and IDC Data- https://www.seagate.com/files/www-content/our-story/trends/files/idc-seagate-dataage-whitepaper.pdf, accessed on 21-06-2019
[5] Statista SC IoT - https://www.statista.com/statistics/423063/smart-cities-connected-things-installed-base-utilities-sector/, accessed on 21-06-2019

4. Structure of the thesis

Introduction

In this chapter, we intend to frame the main elements that support the research work, defining the research problem, the objectives, whether the general objective or the specific objectives, the purpose and justification of the research work, and the revelation of the structure of the research thesis.

Part I - Theoretical framework.

This part presents a review of the literature that covers the state of the art-oriented to the subjects dealt with in this research work, the more generic concept of smart places, smart cities, blockchain technology, and its various applications, the IoT, and more specifically in the mobility and transport ecosystem, in rail transport, ticketing, and traffic management systems.

Data management remains central to this review, both in its conceptualization and in its governance.

The first point presents smart places and the implementation scenario in smart cities. Defines the concepts of smart places and reviews the models that allow the concepts of smart cities to structure natural ecosystems through data ecosystems.

The second point focuses on blockchain technology, from its characteristics, the block structure, the digital signature, the main characteristics of the blockchain, the consensus mechanisms such as PoW (Proof of Work), PoS (Proof of Stake), PBFT (Practical Byzantine Fault Tolerance), dPoS (Delegated Proof of Stake), Ripple (Ripple Protocol Consensus Algorithm), Tendermint, other consensus algorithms, and their main challenges. It also reviews the application aspects in security, blockchain platforms, architectures and taxonomy, the use of Merkel trees, and the IPFS (InterPlanetary File System), as ways to enhance and optimize the use of blockchain technology.

For the adoption of blockchain technology, the way in which blockchain technology can be used as a basis for ensuring data trust, controlling data flows, and supporting information integrity is reviewed.

The third point reviews the main aspects of platforms supporting data markets. Data markets are analyzed in the context of smart cities and reviewing the importance of trust in the data available, in its quality and origin. The fourth point reviews the mobility and transport system that will be the object of artifacts based on proofs of concepts.

From smart places and smart cities models, we focus on the mobility and transport ecosystem, with public rail transport systems, information technologies in public transport networks, with information systems to support control and management, operation, management of rail traffic and rolling stock, and specifically the ticketing system, the central object of the design and development.

Part II - Empirical study.

The first point of part II is dedicated to the methodology, with the definition of the method, the foreseen steps, the problem, the intended development, how to evaluate, and the conclusion. At this point, we intend to structure the research methodology, with the description and characterization of the research and the foundations for the methodological option and the general contextualization of smart places and in particular the smart city under analysis.

In the second point, the research results are presented, with the generic model, the system architecture, and the practical results. In this chapter, we intend to create the contextualization, operationalization, and analysis of the smart city and the specific case under analysis, applied to the mobility and transport system, in the application domain of the ticketing event system.

In the third point, the results are analyzed and discussed. This chapter intends to systematize, develop and evaluate the proposed model in order to be generalized. The analysis and construction of the model to be adopted must allow the operationalization and evaluation of data management, its governance, and flow control, representing a contribution of Value to the smart city and in a generic way to smart places.

Part III - Final Considerations.

In this part, the conclusions and limitations are presented, where the conclusive aspects of the work are described, emphasizing the contributions that this work provides in the context of the management and control of data flows and the limitations that may condition the application of the proposed model.

II. THEORETICAL FRAMEWORK

II. THEORETICAL FRAMEWORK

After Chapter I, introductory, appears in this chapter the theoretical framework where it is intended to review and present the information obtained through the consultation and careful analysis of the literature review with the scientific evidence considered relevant for the understanding of the themes and that can provide the information that scientifically supports the proposed questions.

This review of the literature and the state of the art intends to consolidate the theoretical framework that should support the main aspects of this investigation. Theoretical aspects and concepts related to the analysis model are deepened to allow the understanding and explanation of the problem with the understanding of the intended practice for the empirical work.

The state of the art thus presents the various concepts involved and the documented applications, strategies, methodologies, and techniques that facilitate the organization, information, and development of activities in the empirical part.

1. *Smart places*

At this point, the different perspectives of smart places are analyzed, such as physical spaces, of variable geometry, with underlying organization models and with their virtual and digital translation, which add other valences and characteristics of modeling, simulation, monitoring, participation, and collaboration, management, and governance, which enhance the more effective and efficient adoption of new, measurable and controllable policies.

The smart place concept redefines the spatial dimension and dematerializes natural ecosystems into digital ecosystems, allowing relationships to be established, integrating, and allowing interoperability. The spatial dimension of smart places is variable, depending on the geography and underlying Organizational models. In this concept are the Entities with a physical expression and which Organizations are. That contains the dimensions, people, governance, and technology.

Morandi, Rolando, & Di Vita (2016) redefine spatial and conceptual references. The smart city concept evolves into the concept of a "smart" region with specific characteristics and potential for innovation. The transition from IoT to the Internet of Places allows the integration of physical and digital services, exploring the spatial context of the "smart" region through the adoption of urban functions important for the innovation of the metropolitan region. The urban node concept evolves into the digital urban node concept that supports the development of the Internet of Places in the concept of the scale of urban functions, with a wider spatial scale and with different categories of users such as residents, students, and cities. -users.

The way in which Organizations, cities, towns, and neighborhoods will respond to the changes that smart translates will be essential for the need to manage their physical environments and for what smart means, with the physical attributes that are can give to their virtual equivalents in creating memorable, prosperous, sustainable and true places. (Walters, 2011)

In the dimension of people, in the study of Glaeser (2006) highly skilled entrepreneurs innovate in ways that employ people with similar skills. To this extent, if qualified individuals increase, the wages of the workers around them increase and we will have qualified individuals concentrated in "smart" cities, with a tendency towards greater concentrations of wealth. As a result, local and regional policies have an interest in ensuring that their communities and regions become "smart" and remain "smart". For this, it will be necessary and essential to invest in education and safety, with the aim of contributing to the influx of highly qualified workers, who value safe streets.

In this research work, the smart place dealt with will be the specific case of smart cities, which is detailed in the next point.

1.1. *Smart city*

At this point, the smart city, as the smart place object of this work, introduces different views, definitions, and models of this "smart" space formed by numerous dimensions and factors that seek to be present and reviewed based on the literature review.

In the article (Brandão et al., 2018a) developed within the scope of this thesis, it is considered that the adoption of blockchain technologies, as a platform for governance and for the security and reliability of data, adapted to the different characteristics of the city and that goes through combining IoT objects, various networks, various ecosystems that interconnect, can result in safer and more efficient cities.

Cities start to develop into smart cities more due to corporate needs with marketing campaigns than social "intelligence". (Deakin & Al Waer, 2011). The social "intelligence" to develop depends on the quality and wealth of information and communication in this transition. "Smart" communities play a critical role in cities in creating networks of innovation, creative partnerships and learning, knowledge transfer, and capacity building.

The digitalization process and the electronic infrastructure underlying the service sector allow for cost reductions and the improvement of relationships between buyers and suppliers, which multiply and facilitate negotiations and transactions. (Scuotto et al., 2017). With the sharing of knowledge and the interconnection of more global electronic markets, cities or regional spaces have to leverage themselves on four pillars; human resources specialized in ICTs, knowledge sharing activities, buyer-supplier relationships, and the adoption of electronic markets to enhance regional products and services.

The concepts of the "smart" city and the spatially enabled society presented by Roche, Nabian, Kloeckl, & Ratti (2012) are two different yet related fields of the city. The infrastructure and data community drive the former concept, while practitioners and researchers particularly in urban planning, urban studies, and urban design are more concerned with the latter.

Wolisz, Böse, Harb, Streblow, & Müller (2014) present the city/region information model (CDIM-City District Information Model) as a concept of integrated data management for city neighborhoods, to simulate different qualities of input data, showing that small deviations from building standards or dimensions of construction considerably affect energy requirement and costs.

Spiekermann & Cranor (2009) present the privacy requirements for smart cities, based on historical and contemporary perspectives, through a model with three levels of concerns, with the privacy of related users, with the system operations that go through data transfer, storage, and processing, and with the analysis of user behavior to develop guidelines for building "privacy by design" systems.

Deakin (2011) suggests that the "intelligence" embedded in smart cities makes it possible to activate the creativity of emerging virtual communities of a digital inclusive nature, with analytical and synthetic capabilities to create virtual communities, through the use of collective memory, wikis, and service blogs. Electronically enhanced as a means of bridging social divisions.

Beyer, Elisei, Popovich, Schrenk, & Zeile (2015) present sustainable urban governance through the interactive and participatory process of the citizen and not just users, in the paradigm of "smart" mobility, with the open and active involvement of people and stakeholders, in the areas of technology, transport, land use, urban issues, environment, public health, ecology, engineering, green modes, and public transport.

Prehofer et al. (2010) demonstrated the feasibility of providing "smart" smart space services and implementing a "smart" space prototype, by proposing a web-based framework as a software architecture style for web services (REST-Representational State Transfer) to enable "smart" spaces. Smart" and to support applications spread across multiple devices.

Sánchez, Elicegui, Cuesta, Muñoz, & Lanza (2013) present an architecture exploring the main concepts of the paradigm of the Internet of the Future (FI-Future of the Internet) to create more "smart" cities. The architecture is supported by existing critical communications infrastructure, ownership of public services that allow the integration of current vertical city services and for the efficiency and sustainability of our cities. The prototype was deployed in a park in the city of Santander for the adaptation of autonomous public lighting.

Mohammadi, Al-Fuqaha, Guizani, & Oh (2018) present the Deepen Reinforcement Learning (DRL) in support of IoT and "smart" services, where the "intelligence" of services is obtained and improved through sensory data, with context recognition, achieving that data unlabeled and providing user feedback that allows for learning and choice between alternative actions.

Walters (2011) argues that if urban design implementation policies and tools are embedded and codified within a community's e-governance framework, a balance can be achieved when the physical and virtual domains enhance the uniqueness of specific locations.

Rey-Robert (2009) states that we have the ability to measure and see the exact condition of anything being instrumented: supply chains, healthcare networks, cities, and natural systems. The interconnection of people, systems, and objects allows them to communicate and interact with each other in new ways that can respond to changes quickly and accurately, and get better results by predicting and optimizing for future events.

Trindade et al. (2017) examine the terms, smart city and sustainability, oriented towards the sustainable development of cities. It also presents a theoretical basis for understanding the relationship between the concepts of sustainable urban development and "smart" cities, as a challenge to make cities more attractive to people.

S. Li, Yang, & Gao (2015) study how smart cities were developed in China, especially the roles and relationships of various actors, including government, the market and society in the development of the smart city.

De Jong, Joss, Schraven, Zhan, & Weijnen (2015) investigate the twelve most used categories of dominant cities: sustainable city, ecocity, low carbon city, livable city, green city, smart city, digital city, ubiquitous city, "smart" city, information city, knowledge city, and resilient city, its conceptual differences and the interrelationships between the twelve categories The variety of city categories blend together with the aim of creating social, economic and environmental sustainability or regeneration.

Schaffers et al. (2011) explore the smart city concept as an open and user-oriented innovation environment to experiment and validate new services. The project, in the field of Living Laboratories, is oriented towards common research and innovation resources, with the aim of establishing urban and regional innovation ecosystems through partnerships and sustainable cooperation strategies between the main stakeholders.

Alawadhi & Scholl (2016) present the study that documents the governance models that emerged in four smart city initiatives (Seattle / USA, eCityGov Alliance / USA, Munich / Germany, and Turin / Italy) in which governance depends on a number of different factors, on a case-by-case basis, but where stakeholder involvement in governance was considered critical in all cases.

Kitchin (2014) focuses on how cities are being instrumented, with digital devices and infrastructure, which produce big data. Smart city data allows real-time analysis of city "life", forms of urban governance, and provides the basis for envisioning and promoting a more sustainable, open, transparent, competitive, and efficient city.

Khatoun & Zeadally (2016) reflect on the importance of concepts, architectures, and research opportunities in smart cities given the predictable improvement in the quality of life of its citizens. It focuses on aspects such as IoT management, data management, smart city assessment, security, and renewable technologies, with challenges that require proactive solutions in security and privacy.

Albino, Berardi, & Dangelico (2015) seek to clarify the meaning of the word "smart" in the context of cities and identify the main dimensions and elements that characterize a "smart" city, reviewing the different metrics of urban "intelligence" in the common definition of smart city and its characteristics.

Braem et al. (2016) present in a City of the Things test environment, in the city of Antwerp, Belgium, through the multi-technology network infrastructure, with the ability to test new data and validate it.

Streitz et al. (2005) present work that designs "smart" artifacts in "smart" environments to improve relationships between participants in distributed workgroups. Participants maintain their personal mobility, adding possibilities for collaboration, informal communication, and social awareness that can contribute to sharing and cohesion.

Schleicher, Vogler, Dustdar, & Inzinger (2016) discuss the significant challenges through an overview of SCALE (Smart city Application Ecosystem), supported by middleware such as the smart city operating system. The smart city application ecosystem aims to integrate stakeholders and resources to efficiently build, deploy and operate the smart city's most important applications. It also reveals the ideal "smart" city cycle through a conceptual model for a reactive system that addresses the current challenges of smart city applications.

F. Li, Nucciarelli, Roden, & Graham (2016) present a framework that illustrates new business models and how smart cities redefine scalability, analytics, and connectivity. Smart cities can thus transform operational models and verify the viability, vulnerability, and acceptability of each new operation.

Nam & Pardo (2011) present the conceptualization of the smart city in the dimensions of technology, people, and institutions, through strategic principles such as the integration of infrastructures and technological mediation services, social learning for the consolidation of human infrastructure, governance for institutional improvement, and the involvement of citizens.

Benevolo, Dameri, & D'Auria (2016) analyze "smart" mobility initiatives as part of the smart city, and investigate the role of ICTs in supporting "smart" mobility actions, the impact on citizens' quality of life, and the value created for the city as a whole.

Grave (2016) refers to the essential role of stakeholder engagement in the various smart cities projects. The commitment is based on exploring data and information as knowledge, on adopting technology and innovation as a competitive advantage, on incorporating project planning and methodology, through trans-organizational accountability, on having the collective leadership approach and reinforcing individual change, to promote the trans-informational relationship, to use education for inclusion and to guide sustainability in improving the quality of life.

Dameri (2012) seeks a comprehensive proposal for the definition of the smart city given that the meaning attributed to this expression is still based on empirical experience and not on systemic theoretical studies for the effective implementation of the smart city, capable of creating public value, well-being for citizens, and environmental sustainability. The concept of the "smart" city is used to identify a wide spectrum of heterogeneous solutions and municipal programs, involving different types of technologies and aiming to achieve a very large set of different and not well-defined objectives. These terms are used differently to define similar projects and solutions, attributed to the idea of a smart city.

The definition of the smart city will be based on physical space, citizens, technology, and governance, through digital, sustainability, inclusion, and democracy, with different limits and dimensions, with well-defined and measurable objectives of environmental sustainability, capital creation intellectual intelligence, citizen participation, and well-being.

In this perspective of Dameri, the following definition of smart city is presented:

> "A smart city is a well-defined geographic area, in which technologies, such as ICTs, logistics, energy production, and so on, cooperate to create benefits for citizens in terms of well-being, inclusion, participation, environmental quality, development. "Smart", and is governed by a well-defined set of subjects, capable of defining the rules and policies for the government and the development of the city."

(Dameri, 2012)

Reviewing the different perspectives of the smart city concept, we can define, in short, the smart city as a multidimensional and multifunctional space, seen as a global and "smart" ecosystem, sensed in IoT and organized in functional ecosystems that group application domains, which relate and integrate, as an expansion of spaces and communities, which involves the entire physical and digital space, focused on the citizen, involving participation and feedback, to make the needs of the citizen compatible with sustainable public policies.

1.2. Smart city models

As a result of the processes of systematization and modeling of smart cities, several works have emerged that seek to respond to this challenge of defining models for smart cities.

Dustdar, Nastic, & Skekic (2016) present a vision of the "smart" cyber city, based on the architecture of values, which characterize the complex coordinated activities that involve city services, stakeholders, and "smart" devices, centered on citizens, which promotes participation rather than passivity. In this vision, it defines a set of key enablers such as complex coordinated activities, incentives as flexible control and supply mechanisms, and governance based on the infrastructure of public services. In addition to defining design principles and requirements for cities of the future and how to realize a comprehensive platform.

Colding & Barthel (2017), using an urban ecology perspective, reflect on smart city models with the aim of increasing urban sustainability, social sustainability, health, resilience, and cybersecurity, which can affect the autonomy of urban governance, personal integrity, and the resilience of infrastructure that provides inhabitants with basic needs, and how smart city developments can change man's relationship with nature.

Amorim (2016) presents the city information modeling (CIM-City Information Modeling) where it lists the essential factors for the implementation of the CIM analogous to the BIM (Building Information Modeling) paradigm and complementary to the smart city concept. The resources used by the GIS (Geographic Information Systems) tools, the CityGML (City Geography Markup Language) standard, and the IFC (Industry Foundation Classes) standard allow the identification, projection, simulation, monitoring, and urban management. This CIM model must consider the concepts of planning, design, construction, operation, and maintenance. The conceptual information model must map objects and processes, and the implementation with the encoding of data structures that must be able to represent the different types of objects, their geometries, properties, relationships, and states, through sharing, interoperability, reliability, and security.

Rathore, Ahmad, Paul, & Rho (2016), in the context of urban planning and construction of smart cities based on IoT and big data analysis, propose a 4-layer architecture: Layer 1 - Lower layer: responsible for IoT sources, data generation, and collections; Layer 2 - Intermediate Layer I: responsible for all types of communication between sensors, relays, base stations, Internet, etc.; Layer 3 - Intermediate Layer II: responsible for data management and processing using Hadoop framework; and Layer 4 - Upper layer: responsible for the application and use of data analysis and results obtained. The system consists of data generation, collection, aggregation, filtering, classification, pre-processing, computation, and decision making.

Ghannem, Hamdi, Abdelmoez, & Ammar (2015) present a development process and a modeling approach for the construction of the environmental context model through the RE-CAWAR (Requirement Engineering Context Awareness) methodology, guided by the environmental context model, where the environmental dimension is the dimension with the greatest impact on the changes in a dynamic context.

Abella, Ortiz-de-Urbina-Created, & De-Pablos-Heredero (2017) present a model for the analysis of data-based innovation and value generation in smart cities ecosystems. The model works in three steps: it adjusts the disclosure of data by the smart city with dimensions that build attractive and reusable data, analyzes the mechanisms to create innovative products and services, and explains how these products and services can affect society.

Ceballos & Larios (2016) propose a model to increase citizen participation and support the development of the smart city master plan, adapting the Kano model (Kano et al., 1984) based on KPIs (Key Performance Indicator) (Cohen, 2013) and the ISO 37120 standard (ISO 37120, 2014)to feel the citizens' perception. Table II.1, adapted from table 1, from Ceballos & Larios (2016), presents the comparison between these two models of smart cities metrics. Boyd Cohen's smart cities circle model recognizes six KPIs to classify a smart city and ISO 37120 includes 17 key measures:

Table II.1 - Comparison of two smart cities metrics models

Smart cities circle	ISO 37120
"Smart" economy - *Smart Economy*	Economy
	Finance
"Smart" environment - *Smart Environment*	Energy
	Environment
	Solid waste

Smart cities circle	ISO 37120
	Residual waters
	Water and Toilets
"Smart" living *Smart Living*	Fire and Emergency Response
	Health
	Safety
	Shelter/housing
"Smart" Mobility - *Smart Mobility*	Telecommunications and Innovation
	Transport
	Urban planning
"Smart" people - *Smart People*	Education
	Leisure
"Smart" Government – *Smart Government*	Governance

Adapted from table 1, from Ceballos & Larios (2016)

Wei, Huang, Li, & Xie (2016) present an Urban Carrying Capacity (UCC-Urban Carrying Capacity) assessment model as a basis for improving urban sustainability. The integrated analytical framework of UCC, Table II., adapted from table 5 of Wei, Huang, Li, & Xie (2016), of 30 indicators representative of the literature, is presented below and was applied to megacities in China, with the indicated weights (Weight1, Peso2, and Peso3):

Table II.2 - System of indicators by system and evaluation area, weighted

Sector	weight1	Evaluation areas	weight2	indicators	weight3
economic	17.60%	Job	3.40%	X1 Urban registered unemployment rate (%)	3.40%
		Affluence	7.00%	X2-Disposable income of urban families per capita	3.70%
				X3-Tax Revenue per capita (Euro)	3.30%
		Economic Scale	3.80%	X4-GDP per capita (Euro)	3.80%
		Growth	3.40%	X5- Annual GDP growth rate	3.40%
Resources	17.90%	Water	7.40%	X6-Per capita water supply (ton)	3.00%
				X7-Domestic consumption of daily water per capita (liters)	4.40%
		Earth	3.70%	X8-Constructive land per capita (m2)	3.70%
		Energy	6.80%	X9-Per capita gas supply (m3)	3.10%
				X10-National consumption of electricity per capita (kWh)	3.70%
Environmental	25.80%	Pollution	6.80%	X11-Discharged of industrial waste water for 10 thousand Euros of GDP (ton)	3.90%
				X12-Industrial CO2 emissions per 10 thousand Euros of GDP (kg)	2.90%
		Treatment	13.30%	X13-The proportion of solid industrial waste that is used extensively	4.00%
				X14- The proportion of treated wastewater (%)	3.40%
				X15- Urban waste treatment fee	2.90%
				X16-The number of days with air quality above standard class 2 per year	3.00%

Sector	weight1	Evaluation areas	weight2	indicators	weight3
Infrastructure	38.60%	Green space	5.70%	X17-Green area per capita (m2)	2.50%
				X18-Green coverage rate of urban built-up areas (%)	3.20%
		Health	2.90%	X19-Number of hospital beds per 10,000 people	2.90%
		Housing	3.40%	X20-Per capita space of urban residents (m2)	3.40%
		Utilities	10.40%	X21-The density of the drainage pipe in urban areas (km / km2)	2.90%
				X22-T+I6water access rate (%)+I21	3.70%
				X23-Gas access fee (%)	3.80%
		Communication	9.20%	X24-Internet number per 10,000 people (user)	3.00%
				X25-Number of mobile phone users per 10,000 people (user)	2.70%
				X26-Number of fixed telephony users per 10,000 people (user)	3.50%
		Transport	12.70%	X27-Bus number per 10,000 people (unit)	3.10%
				X28-Number of private cars per 10,000 people (unit)	3.10%
				X29- Areas of urban roads per capita (m2)	3.60%
				X30-Density of roads (km / km2)	2.90%

Adapted from table 5, from Wei, Huang, Li, & Xie (2016)

Chilipirea, Petre, Groza, Dobre, & Pop (2017) present an integrated architecture oriented to the flow of data, from the origin to the end-user, for smart cities, with data processing steps, with the collection of heterogeneous sources, normalization, intermediation, storage, analysis, visualization, and the delivery to services or applications or decision support systems.

Lee, Phaal, & Lee (2013) present a roadmap, applying the QFD (Quality Function Deployment) method, to establish interconnections between services and devices, and between devices and technologies, integrating services, devices, and technologies capable of implementing a "smart" city development project. Table II.2, adapted from table 8, Lee, Phaal, & Lee (2013) suggests detection, network processing, interface and security categories.

Table II.2 - Smart city technology rating standard

Category	Definition
Detection	Monitor any external state changes and transmit the collected data to process and respond to sensor signals
Processing	Treat process data from sensors according to an analysis, leading to a rational decision
Network	Connect each device and user to support efficient communication
Interface	Convert the information that flows between devices or between users and devices into a more intelligible form (graphic, structural)
Safety	Control illegal access to user or facility information throughout the "smart" environment and protect personal privacy

Adapted from table 8, from Lee, Phaal, & Lee (2013)

Badii et al. (2017) concluded that smart city solutions tend to transform data into services, for city users and operators, through the exploration of data solutions and data analysis whose services integrate open and private data, static and real-time data from public entities and private operators. The smart city APIs (Application Program Interfaces), depending on the chosen architectural solutions, to pass from data to services, allow different functionalities, such as the exploration of aggregated data and processed by algorithms, for the production of services.

Table II.3, summarized in table 1, of Badii et al. (2017), for different "smart" city solutions, to move from data to services can be applied to the cases of information integrator, data and metadata aggregator and semantic aggregator and reasoner, limited to data interoperability between city entities, in: time, space, multiple domains (interoperable semantics), structures, services, and relationships.

Table II.3 - "Smart" city solutions to move from data to services.

Smart cities solutions to move from data to services
Addressing open data (open data)
Addressing private data
Addressing data in real time
Addressing interoperable services
Pushing data collection
Collecting data by pulling (Pull)
Providing data lookup
Providing metadata lookup
Providing spatial reasoning
Providing time reasoning
Providing integrated authenticated access to data
Providing syntactic interoperable data/services
Providing semantic interoperable data/services
API independent of data model changes
Providing REST API on data
Providing SPARQL API on Data
Providing inference support on data
Providing data visualization (business intelligence, dashboarding)
Providing decision-making support

Adapted from table 1, from Badii et al. (2017)

The Table II.4, summarized in table 3, of Badii et al. (2017) presents the front-end domains of smart cities APIs to provide services for "smart" city management applications, and for web and mobile applications.

Table II.4 - Smart city API solutions

Smart city API frontend domains
API: Search Service
(GPS API) full text search
(GPS API) Search around a GPS point
(GPS API) Get GPS location
(Location API) Search along a line, a polygon
(Location API) Search in an area, a closed shape
(Location API) Search for street, region, county, etc.
API: Mobility
Get Public Transport, bus stops, lines, and schedule
Get real time delay of hi public buses from other modes of transport
Get fluid traffic status
Get parking status
Get price status at gas stations
Get vehicle location and route (Current: Latitude, Longitude | POI; Destination: Latitude, Longitude | POI)
Get an intermodal referral
Get an integrated ticketing
Getting a route to a good delivery (multi-stop planning)
API: Environment, sensors and actuators, IoT, healthcare
Obtain healthcare facilities (hospitals, doctors, etc.)
Get first aid status
get forecast
Get Sensor / Actuator Value / Status
Get pollution, temperature, pollination, etc.
API: User participation and awareness
Get information about social media monitoring
Save comments provided by the crowd (crowd sourcing) by service
Save the votes and average provided by the crowd (crowd sourcing) per service
Get/Set variable message panel state by location
(API EVENT) Get city/area events (today, week and month)
API: Personal assistant (appointment + recommendation)
Save User Profile
Get tailor-made suggestions (on demand)
Get sound information, information, appointments
Receive civil protection news (in push)
Save the state of mobile sensors
API: Georeferenced Service Domains
API: Call Type (JSON and/or HTML)
SPARQL Query (SPARQL Query)
SPARQL Query with Inference (SPARQL Query with Inference)
REST
Query ID (Query ID)
API for non-functional features
API direct authentication
API authentication via social media
Data licensing control

Adapted from table 3, from Badii et al. (2017)

Bellini, Nesi, & Pantaleo (2015) reported a comparative study of benchmarking repositories of information modeling and description frameworks (RDF-Resource Description Framework) for smart cities services based on their main features of SPARQL[6]. The RDF[7] is a framework for modeling and describing information, labeled and directed to represent information on the Web. The RDF store[8] analyzed were the following: Virtuoso[9], GraphDB[10], Blazegraph[11], CumulusRDF[12], Stardog[13], and the Strabon[14]. The smart city RDF benchmark proposal[15] based on Florence smart city accessible as Km4City[16]. The evaluation model aimed to validate whether RDF stores are suitable for smart city modeling and its application. RDF stores can be used to integrate data from various sources and applications use this data to provide new services to citizens and public management.

Anthopoulos, Janssen, & Weerakkody (2015) compared smart cities with different models, through the systematization of modeling and benchmarking methods. They established the common dimensions, people, government, economy, mobility, environment and experience.

The Table II.5, adapted from table 1 of Anthopoulos, Janssen, & Weerakkody (2015) presents the different approaches to modeling smart cities:

Table II.5 - Modeling approaches for smart cities

Source	Model	Description
(Söderström et al., 2014)	Nine pillar models	Planning, infrastructure service management services and human services
	Smart city equation that combines instrumentation, interconnection and "intelligence".	Instrumentation (the transformation of urban phenomena into data) + interconnection (of data) + "intelligence" (brought about by the software)
(ITU-T, 2014)	*smart city*: Key Performance and Sustainability Indicators	Environmental sustainability, productivity, quality of life, equity and social inclusion and infrastructure development
(UN-Habitat, 2013)	Dimensions of City Prosperity	Productivity and prosperity of cities, urban infrastructure: foundation of prosperity, quality of life and urban prosperity, equity and prosperity of cities, environmental sustainability and prosperity of cities
(Anthopoulos, 2015)	Smart city dimensions	Resources, transport, urban infrastructure, well-being, government, economy and coherence
(ISO 37120, 2014)	ISO 37120 sustainable development of communities, indicators of city services and quality of life	Economics, education, energy, environment, finance, fire and emergency response, governance, health, leisure, security, housing, solid waste, telecommunications and innovation, transport, urban planning, wastewater, water and sanitation.

---

[6] SPARQL - http://www.w3.org/TR/sparql11-query/ , accessed on 2019-03-12
[7] RDF - https://www.w3.org/RDF/ , accessed on 2019-03-12
[8] RDF store - https://db-engines.com/en/article/RDF+Stores , accessed on 2019-03-12
[9] Virtuoso - https://virtuoso.openlinksw.com/rdf-quad-store/ , accessed on 03-13-2019
[10] GraphDB - http://graphdb.ontotext.com/, accessed on 03-13-2019
[11] Blazegraph - https://www.blazegraph.com/, accessed on 03-13-2019
[12] CumulusRDF - https://www.w3.org/2001/sw/wiki/CumulusRDF, accessed on 03-13-2019
[13] Stardog - https://www.stardog.com/, accessed on 03-13-2019
[14] Strabon - http://www.strabon.di.uoa.gr/, accessed on 03-13-2019
[15] smart city RDF - http://www.disit.org/smartcityrdfbenchmark, accessed on 2019-03-14
[16] Km4City - https://www.km4city.org/?devTools , accessed on 2019-02-20

Source	Model	Description
(Neirotti et al., 2014)	Smart city domains	Natural resources and energy, transport and mobility, buildings, well-being, government, economy and people
BID[17]	Emerging and Sustainable Cities Initiative (ICES) Application of the economic filter for prioritizing action areas	Qualitative economic impact decision method. *Urban Dashboard* [18] - Urban Dashboard that allows you to explore and compare more than 150 quantitative indicators
(Lee et al., 2014)	Smart city analytics board	Urban openness, service innovation, partnerships, training, urban pro-activity, integration of "smart" infrastructure in the city, "smart" city governance

Adapted from table 1, from Anthopoulos, Janssen, & Weerakkody (2015)

Table II.6, adapted from table 2 of Anthopoulos, Janssen, & Weerakkody (2015), presents some of the smart cities assessment tools.

Table II.6 - Smart city benchmarking tools

Source	Benchmarking tool	Description
(Moreno Pires et al., 2014)	Local Sustainable Development Indicators	21 ECOXXI indicators, grouped in the following sectors: sustainable, education, development, sea and coast, environmental institutions, nature and biodiversity conservation, forestry planning, air, water, waste, energy, transport, noise, agriculture and tourism
(Kourtit et al., 2014)	*Global City*: Performance measurement indices	Economy, research and development, cultural interaction, quality of life, environment and accessibility
(Desouza & Flanery, 2013)	City Resilience Assessment and Implementation Framework	City Components: Resources and processes (physical) people, institutions and activities (social)
(Cross & Marques, 2014)	*scorecard*: Sustainable Local Government	Social, economic, environmental and governmental criteria
(Singhal et al., 2013)	Competitiveness parameters	Physical environment, social capital, finance, development, investment, user potential
UN Habitat (UGI, 2004)	Urban governance indicators	Effectiveness, equity, participation, accountability and security
(Lazaroiu & Roscia, 2012)	Model to calculate smart city indices	Economy, mobility, environment, people, well-being, governance
(Duarte et al., 2014)	Digital City Assessment Framework	Connectivity, accessibility and communicability

Adapted from table 2 of Anthopoulos, Janssen, & Weerakkody (2015)

The various modeling approaches and assessment tools for smart cities raise several questions when it is intended to systematize and identify repeatable standard tests and on which data should be supported. Which analytics provide Value? What is the most suitable data? How to identify priorities? How to maximize Value for the city? How to leverage best practices in the city? How to deploy technology-adjusted to its size? The most "smart" cities should be cities that anticipate problems to solve them in a proactive and coordinated way and that take advantage of the most adjusted information to make

---

[17] BID - https://www.iadb.org/es/desarrollo-urbano-y-vivienda/programa-ciudades-emergentes-y-sostenibles
[18] Urban Dashboard - http://www.urbandashboard.org/iadb/index.html , accessed on 4-04-2019

better decisions, with the aim of stimulating sustainable economic growth. (McNamee, 2009).

The fundamental concepts of the design of a "smart" city are extensively revised and with numerous initiatives. Security vulnerabilities, privacy issues in the context of smart cities and their services maintain the need for deepening and investigation. ICTs and city infrastructure enhance the city's objective of improving the quality of services it provides to citizens and improving their quality of life. (Khatoun & Zeadally, 2016). Table II.7, adapted from table 2, from Khatoun & Zeadally (2016) describes the risk-oriented cybersecurity features of the built environment.

Table II.7 - Security standards and norms and recommendations for smart places cybersecurity

Feature	Description	Standards and recommendations
Organizational	Develop a backup and recovery plan	Five Best Practices for Improving Building Management Systems, Schneider Electric(Strass & Williamson, 2014)
	Manage users and passwords	EIT Standards: - Cybersecurity risks in the built environment - Standards, skills and learning (Boyes, 2016) - "Smart" Buildings: Understanding and managing security risks (IET, 2012)
	Open feedback sessions	Cybersecurity in "smart" buildings (Cybersecurity in Smart Buildings, 2015)
	Define standards, tools, security procedures and rules for the community	Measuring Science Roadmap for Net-Zero Energy Buildings(Pellegrino et al., 2010) Understanding risk and investing in resilient infrastructure (Gallego-Lopez & Essex, 2016)
	Develop policies related to password and configuration management	Information Sharing and Analysis Centers (ISACs) - Cooperative Models(ENISA, 2017) Smart city standard concept model (ISO 30182: 2017) Standard on existing guidelines and approaches on sustainable development and resilience in cities (ISO 37121: 2017)
Technician	Provide physical security for equipment, network cables and servers Network traffic and encrypt with robust symmetric algorithms such as AES and Blowfish[19]	ANSI/TIA-862, Building, Automation Systems Standard Cabling GSA Guide Specification Interoperable Building Automation and Control Systems Using ANSI/ASHRAE 135-1995, BACnet
	Use of secure connection such as a VPN for remote access	• Security requirements of "smart" buildings based on the Internet of Things using RESTful web services (Niemeyer et al., 2014)
	Secure wireless network with WPA2	• Federal Agency BSI[20] Information

---

[19] blowfish- https://www.schneier.com/academic/blowfish/ , accessed on 5-04-2019
[20] BSI - https://www.bsi.bund.de/EN/Home/home_node.html , accessed on 5-04-2019

Feature	Description	Standards and recommendations
	protocol *IDS Deploy* in construction Use centralized authorization and accounting (AAA) server authentication as a RADIUS server Deploy a firewall at every transition point Deploy a firewall at every transition point Deploy a firewall at every transition point Using strong authentication methods such as biometric or "smart" cards	Security
Human Resources	Comprehensive training program for programmers and system administrators. Inform and raise awareness of security issues Alerting and advising key users where threats exist Continuity plans to ensure disaster recovery	ISO 27001 Framework Cyberinsurance Citizen[21] Intrusion Detection Recommendation UC Berkeley security policy-guideline[22]) ISO 22301 - Business continuity
Nice	Respect legal aspects of security Use security standards and recommendations from national cybersecurity, agencies and IT security actors for follow-up Good practices in the use of ICT control and performance Personal data protection	Law No. 46/2018, of 13 August - Cyberspace Security Legal Regime ISO/IEC 27001 - Standard for information security management system NIS Directive (Network and Information Security Directive) on network and information security (EU) 2016/1148 ITIL - Best Practices (Alter, 2015) COBIT - Information Control Objectives and Related Technologies Regulation (EU) 2016/679 of the European Parliament and of the Council, General Data Protection Regulation (GDPR)

Adapted from table 2, from Khatoun & Zeadally (2016)

Smart cities, as large consumers and producers of data, present data security challenges. Connectivity transforms smart cities into complex environments where traditional security analysis is not enough and it is necessary to add specific data security requirements and new solutions. The four-layer framework presented by Popescul & Radu (2016) lists the critical elements for the functioning of the smart city: "smart" things, "smart" spaces, "smart" systems and "smart" citizens.

Data-driven ecosystems supported by IoT need reliable data collection as a result of the communication and interaction of natural ecosystems. Analytics is a fundamental

---

[21] CNCS - https://lms.nau.edu.pt/courses/course-v1:CNCS+CC101+2018_T1/about , accessed on 5-07-2019
[22] UC Berkeley security policy - https://security.berkeley.edu/intrusion-detection-guideline , accessed on 5-04-2019

prerequisite for urban "intelligence". The smart city depends on the ability to translate IoT data into useful and innovative services. (Dhungana et al., 2016).

The Figure II.1, adapted from figure 1, from Dhungana et al. (2016), translates the smart city as dynamic data-driven ecosystems.

Figure II.1 - Overview of the Data-driven Ecosystem and its components

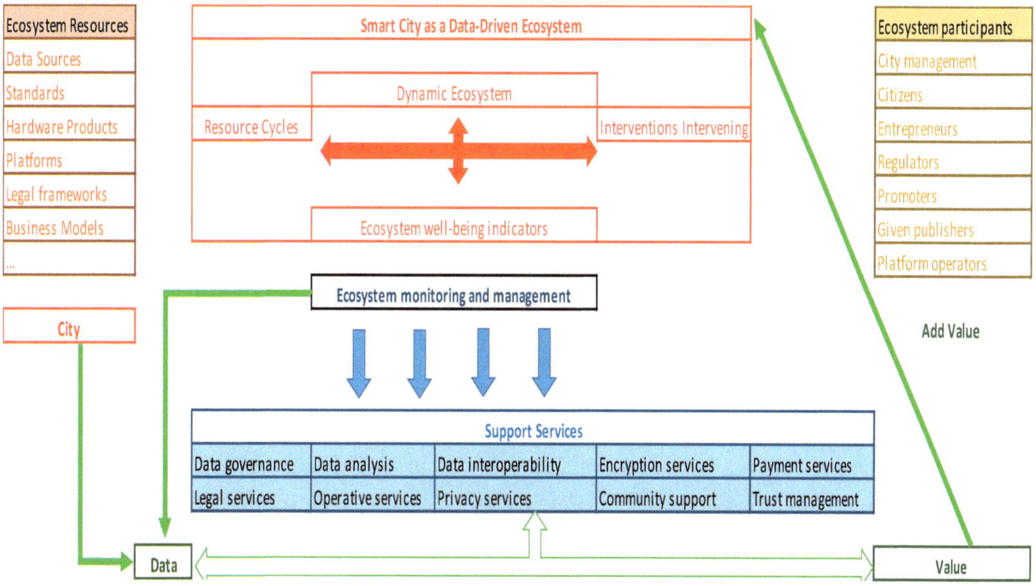

Adapted from figure 1, from Dhungana et al. (2016)

This figure summarizes the resources, participants, and services to support the data-driven smart city ecosystem. Monitoring and managing the ecosystem allows the city, through data, to add Value, which "feeds" the cycles of resources and the interaction of smart city stakeholders, in a dynamic and measurable way through ecosystem well-being indicators.

Cities consume 75% of the world's energy production and generate 80% of CO_2 emissions. The model to calculate the "smart city" indices reveals that the chosen indicators are not homogeneous and have different weights. The indicators are those considered in Table II.8, adapted from table 1, from Lazaroiu & Roscia (2012), for a broader understanding and simple use, for the process of policy formulation and adoption measures and their evaluation. (Lazaroiu & Roscia, 2012)

Table II.8 - Smart city indicators

List of Indicators
1. Pollution.
2. Innovative capabilities
3. CO_2

List of Indicators
4. Transparent governance
5. Sustainable resource management
6. Waste separation
7. Educational facilities
8. Health Conditions.
9. Sustainable, innovative and safe public transport
10. Pedestrian areas
11. Bike paths
12. Green areas
13. Production of solid urban waste
14. Domestic GWh
15. Fuels
16. Policy strategies and perspectives
17. Availability of ICT infrastructures
18. The flexibility of the labor market

Adapted from table 1, from Lazaroiu & Roscia (2012)

This table makes it possible to identify the list of quantifiable indicators that can provide historical evolution and quantify the impact of interventions carried out in this space, such as changes in policies, public investment, environmental and social regulations, among others.

The economic dynamics of smart cities are developed in a model of the dynamics of systems of a city, as a complex and adaptive system of systems unifying the approaches to the Gross Domestic Product of the City, through a "system thinking" for decision-makers, politicians. (Hennessy et al., 2011)

The use of algorithms, designed to exploit the knowledge of the combined agents, which need to cooperate and integrate their information, to provide advanced services for smart city users and to guarantee the interoperability of the different agents, need to consolidate different and heterogeneous ontologies that are unified through corresponding ontology techniques, to improve results and provide the most accurate results possible. (Otero-Cerdeira et al., 2014)

The role of academia, as a smart place inserted in smart cities, is also highlighted in the work of Coccoli, Maresca, Stanganelli, & Guercio (2015) which present a more "smart" university model based on a smart city model, through a case study of collaboration between industry and the university, with "smart" solutions to improve the effectiveness of higher education.

In summary, smart city models multiply and complement each other, having as common aspects the need to subdivide the smart city into ecosystems that interact and integrate, to implement weighted evaluation indicators, and to consider people, institutions, and technology.

This perspective of data-oriented ecosystems, which group functions and application domains, reflects the understanding resulting from this research, on the vision of smart places, smart cities, as physical spaces that dematerialize into data ecosystems, grouping application domains with characteristics of interdependence and similar data groups, submitted to the governance of an Organization, which manages, controls and defines policies focused on the well-being of the citizen and on environmental, social and economic sustainability.

1.3. *Smart cities* and IoT (Internet of Things)

The IoT is central to the development of smart cities, and it can even be said that without the IoT there would not be a city that "knows", that "learns", that "controls" itself, and that allows to sustainably change its policies by adjusting its own policies, practices, collaboration, participation to obtain a better quality of life, more development and social integration.

The benefits of combining blockchain with IoT is presented by Mar (2018), based on the idea of building "smart" machines capable of communicating and operating through blockchain, can solve four issues: supervision, the use of cryptography and distributed storage, the smart contract facilities and the general security of the IoT environment.

The smart city through the IoT allows to consolidate imagined urban and universal governance, and "smart" urban areas, seeking to integrate social networks with IoT solutions, innovate with green IoT technologies and IoT middleware solutions with context knowledge, and use artificial intelligence to recreate and combine IoT with cloud computing. (Mali & Kanwade, 2016)

Harrison & Donnelly (2011) refer that the instrumentation of smart cities is a key aspect in the new theories about cities, which allows considering theoretical frameworks at the level of individual actions and not depending on statistical abstractions to understand what is happening. This vision is multidisciplinary and contributes to the architecture, planning, engineering, construction, operation, and governance of cities, which "makes the invisible visible".

The growing and large-scale deployment of IoT technologies in the smart city seeks to obtain monitored operations to make the city more efficient and improve the quality of life of the city's inhabitants. This objective must be based on a secure IoT architecture to prevent cyber-attacks that can compromise the city's main functions, improperly access personal data, and cause serious damage. The architecture presented by Chakrabarty & Engels (2016) contains four basic IoT building blocks: the black network (Black Networks) which encrypts the payload and metadata within an IoT protocol link layer communication, Trusted SDN (Software Defined Networking) controllers that manage and orchestrate the communication flow between IoT nodes and the rest of the network infrastructure; the unified registry to consolidate heterogeneous technologies; and KMS- Key Management System for generating, distributing, storing, revoking, altering and using keys.

Petrolo, Loscri, & Mitton (2017) present the vision of the smart city as a cloud of things (CoT-Cloud of Things) through cloud sensors as a bridge to IoT. Things are the main requirements for the integration of different IoT ecosystems within the Cloud.

Chowdhary & Deep Kaur (2016) characterize the IoV (Internet of Vehicles) mobility models, as part of the IoT, in a smart city. To this end, it simulates VANETs (Veicular Adhoc Networks), with different configuration parameters, creating several scenarios close to the real movements of the nodes, under various heterogeneous mobility models.

From another perspective, a smart home that can multiply in the urban environment is analyzed, in which the IoT, with its applications in the very close and individual environment of people's and families' lives, analyzes a connected home that interconnects digital devices between itself and through the internet, with communications activated by different protocols, being necessary to guide for connectivity standards in the network of the "smart" house. (Samuel, 2016)

Nitti, Piloni, Giusto, & Popescu (2017) present an IoT architecture for a sustainable tourism application in a city with a "smart" environment, in which the IoT is the technological key for the development of "smart" urban environments using aggregated data, integrated into a single decision platform. The architecture intends to optimize the movement of tourist cruise ships in the city of Cagliari, Italy, using transport information, queue times, points of interest, real transport data and an optimization algorithm.

IoT architectures are preferentially service-centric and are beginning to be user-centric, with mobile devices. User-centric architectures, for end-users, make the city an "intelligent body" supported by the Internet to provide services, where user privacy and security issues are central. (Shaikh et al., 2016)

The smart city approach, with sensors and actuators on a dynamic communication and performance infrastructure, which involves the integration and correlation of data and information from different sources and areas, can optimize active management with the use of a heterogeneous network, with various technologies, supported on a modular platform and for different data flows. (Rinaldi et al., 2017)

The use of an IoT infrastructure to create a "living laboratory" can promote energy savings and environmental sustainability and reveals challenges and opportunities for the development of "smart" buildings and cities, in compliance with health and safety expectations or with organizational norms and policies. (Bates & Friday, 2017)

Psomakelis et al. (2016) present an SOA (Service Oriented Architecture) service-oriented architecture, on the RADICAL platform, for the retrieval and analysis of large data sets from social networking and IoT sites, collected by smart city applications and by data aggregation services, through an innovative smart city infrastructure that seamlessly aggregates and combines social big data and IoT to enable efficient "feeling" analysis techniques to reduce registration, retrieval, update and processing time and use storage techniques for frequency representation, in the context of sentiment analysis in big data and for the best algorithmic approach to dimensional mapping.

The goal of the IoT-based smart city is to design and deploy "smart" city solutions that are replicable, scalable and sustainable, and the adoption of a consensus framework. Another objective of future smart cities projects will be to have a distributed IoT communications network, as an infrastructure for the deployment of shareable and replicable smart city solutions. (Rhee, 2016)

Latre et al. (2016) present to the City of Things an integrated and multi-technology testing laboratory, which allows the configuration and validation of new experiences in terms of technology and user. The City of Things is an integrated approach, allowing experimentation at three different layers: networks, data and a live laboratory supported on a wireless multi-technology network infrastructure.

The city tries to be more "smart" and seeks to be safe and perceptive, to optimize resources and improve the quality of life of citizens. The connection of the virtual to the physical world tends to lead real-time services, based on IoT technology, to modify the responses to real situations. (Sonawane & Shaikh, 2017)

The growing amount of data generated in smart cities brings the need for new practices and techniques for effective data management and analysis, and for generating information that can help to use resources in a "smart" and effective way. Rao & Syamala (2017) present some of the personalized services in a smart city environment through semantic modeling (Uceda-Sosa et al., 2011) and the Dempster-Shafer theory. The Dempster-Shafer theory is a decision theory merging various evidence and sources. (Luo et al., 2018)

The most exciting activities mainly involve integrating IoT services and efficiently processing big data for decision making. In this domain, the work of Rathore, Paul, Ahmad, & Jeon (2017) presents a complete system, with various types of "smart" systems, "smart" house, vehicular networks, water system, "smart" parking, and surveillance objects, based on IoT, for planning the future super city, using big data analytics. Support architecture includes four layers and uses the Hadoop ecosystem[23], with MapReduce programming, where the transfer and computation is distributed reliably and scalable.

The research carried out by Arasteh et al. (2016) describes IoT technologies for "smart" cities, the main components and characteristics of a smart city and reveals that the IoT platform with other autonomous and "smart" systems can provide "smart" and wide-spread and pervasive applications, which will be one of the future trends. It also presents the main challenges that go through security and privacy, heterogeneity, reliability, large scale, legal and social aspects, big data, sensor networks and barriers to response to requests (DR-Demand Response).

In the Internet of Things (IoT), the "things" connected to the Internet provide data entry and resources that offer unlimited possibilities for applications and services. IoT-supported smart city systems (Giang et al., 2016), also explored the application development process from a coordination-based perspective through a distributed

---

[23] Hadoop - https://hadoop.apache.org/ , accessed on 2019-04-20

coordination model, which oversees the distributed components for building IoT-based smart city applications.

Bharadwaj, Rego, & Chowdhury (2016) present a solid waste management system from the city of Bengaluru, India, as an IoT-based architectural solution for the efficient automation of the solid waste monitoring, collection and management process. The sensors collect data from rubbish bins sent to the cloud via the Internet using the MQTT (Message Queue Telemetry Transport) protocol.

1.3.1. IoT communication protocols

The standard protocols reviewed at this point are used in IoT and Industry 4.0, and go through the OPC UA protocols.[24], ROS[25], DDS[26], MQTT[27] (Profanter et al., 2019), MQTT-SN[28] (Stanford-Clark & Truong, 1999), AMQP[29] and CoAP[30] (Corak et al., 2018).

The analyzed protocols are presented below with their main characteristics and in a comparative way in Table II.9.

OPC UA is an M2M (Machine to Machine) communication protocol for industrial automation, ROS is a robotics middleware with a set of software frameworks for development, DDS is a middleware protocol and an API standard for data, MQTT offers flexibility in communication standards and functions as a conduit for binary data, MQTT-SN, close to MQTT, is adapted to resource-limited and low-bandwidth devices, AMQP is message-oriented and forwarding reliably and securely, and CoAP is designed for web interoperability and designed for M2M applications.

The Table II.9, adapted from table 1, from Profanter, Tekat, Dorofeev, & Rickert (2019) summarizes some of the features of the standard protocols used in IoT and Industry 4.0. In addition to the analyzed protocols, XMPP is still used[31] (Extensible Messaging and Presence Protocol) which is an open, extensible protocol based on XML (Extensible

---

[24] OPC UA - https://opcfoundation.org/about/opc-technologies/opc-ua/ , accessed on 13-05-2019.
[25] ROS - http://www.openrobotics.org/ , accessed on 13-05-2019.
[26] DDS - https://www.dds-foundation.org/ , accessed on 13-05-2019.
[27] MQTT - http://mqtt.org/ , accessed on 14-05-2019.
[28] MQTT-SN - http://www.mqtt.org/new/wp-content/uploads/2009/06/MQTT-SN_spec_v1.2.pdf , accessed on 14-05-2019.
[29] AMQP - https://www.amqp.org/ , accessed on 05-10-2019.
[30] CoAP - https://tools.ietf.org/html/rfc7252 , accessed on 05-10-2019.
[31] XMPP - https://xmpp.org/ accessed on 05-10-2019.

Markup Language) through gateways and the WebSocket[32] (IETF - RFC 8441) that allows interaction between a web browser and a web server with low load (Corak et al., 2018).

Table II.9 - Comparison of standard protocols used in IoT and industry 4.0

Protocol	OPC UA	ROS	DDS	MQTT	MQTT-SN	AMQP	CoAP
Descriptive	Open Platform Communications Unified Architecture	Robot Operating System	Data Distribution Service	Message Queuing Telemetry Transport	MQTT for Sensor networks	Advanced Message Queuing Protocol	Constrained Application Protocol
Communication	TCP/IP, UDP	TCP/IP, UDP	TCP/IP, UDP	TCP/IP	UDP, not IP	TCP/IP	UDP
Standards	RPC, Pub/Sub	RPC, Pub/Sub	(RPC),Pub/Sub	Pub/Sub	(RPC),Pub/Sub	Pub/Sub	Pub/Sub
Service quality	No	No	Yea	Yea	No	Yea	No
Authentication	User, PKI	(Mac)	PKI	User, PKI	PKI	User, PKI	PKI
SSL/TLS	Yea	No	Yea	Yea	No	Yea	No
Std API	No	No	Yea	No	No	Yea	No
Standards				ISO/IEC 20922:2016 (MQTT) v3.1.1	ISO/IEC 19464 (AMQP) v1.0OASIS (AMQP) V1.0		IETF - RFC7252
Software	open62541	ROS Melodic Morenia	eProsima Fast RTPSOpendds	Eclipse Paho. Eclipse Mosquitto, HiveMQ CE.Rabbit MQ		Rabbit MQ	coap.mecontiki-os

Adapted from table 1, from Profanter, Tekat, Dorofeev, & Rickert (2019)

This table also summarizes some of the most used platforms to allow multiprotocol communication in IoT environments, such as Rabbit MQ[33], message broker, which supports multi-message protocols, delivery acknowledgment, flexible forwarding to queues, and multiple exchanges.

1.4. *Big data*

Big data and open data create new opportunities and new insights into how citizens perceive their "smart" city.

The smart city also reveals itself with the proliferation of various smart aspects of the city, smart lighting, smart waste management, smart parking, smart traffic management,

---

[32] WebSocket - https://tools.ietf.org/html/rfc8441 , accessed on 13-05-2019.
[33] RabbitMQ - https://www.rabbitmq.com/ , accessed on 13-05-2019.

smart energy-efficient buildings and with the emergence of smart devices and smart technology, mainly of connectivity and communication.

These concepts are oriented towards increasing efficiency and improving services, which through IoT recreate applications for "smart" cities. The data generated requires platforms to process, aggregate and interpret the data produced by "smart" devices and technologies.

The use of a big data platform for "smart" buildings, with advanced building management systems (BMS-Building Management Systems), interconnected to various sensors and actuators and with dedicated networks, allows for the enhancement of data analysis and the development of applications, easing the ability to resize the behavior of the "smart" building, meeting the requirements of scalability, data processing, flexibility, interoperability and privacy. This perspective makes it possible to apply automation rules to preserve or increase comfort and save energy. (Linder et al., 2017)

The self-starting process (bootstrapping) of smart cities, through a self-sustaining model, is based on data flows in big data and on the basic procedure to explore large amounts of data through the store API concept. The objective is to dissociate the political element from the technological maintenance of the city, and that in a coherent way a rollout is developed through the city, in phases, in which in the initial phase, utility and revenue are generated, and in the following phases only the service utility is supported and in the last phase the fun/leisure dimension emerges. (Vilajosana et al., 2013)

Moreno et al. (2017) present two applicability scenarios of big data techniques for smart cities where they show the potential of the applicability of this type of techniques to provide profitable services, such as energy consumption and comfort management in "smart" buildings and the detection of travel profiles in the "smart" transport.

Al Nuaimi, Al Neyadi, Mohamed, & Al-Jaroodi (2015) review big data applications to support smart cities and explore the opportunities, challenges and benefits of embedding applications. They also try to identify the requirements that support the implementation of big data applications in services.

The **Erro! A origem da referência não foi encontrada.**, adapted from figure 1, from Khan, Anjum, & Kiani (2013), presents the context of data management and analysis in a Cloud environment.

Erro! A origem da referência não foi encontrada. - Context of data management and analysis of a smart city, Cloud environment

Adapted from figure 1, from Khan, Anjum, & Kiani (2013)

This figure identifies the five smart city ecosystems: smart people, smart economy, smart governance, smart environment and smart mobility. These ecosystems group applications or application services in the cloud, for the acquisition and storage of data, information processing and decision making. The context of data management and analysis in big data will focus on the application needs underlying the policies applied.

The big data analytics platform combines batch and real-time data processing techniques and requires the use of learning algorithms over datasets that are characterized by their large, dynamic and fast nature. In the application, in smart grid, the requirements demand a framework with MapReduce, Stream and Iterative computational power, using Apache Spark as an integrated platform that combines the requirements of batch, real-time and iterative data processing, which allows providing methodologies advanced analysis and machine learning for the electrical grid (Shyam R. et al., 2015).

"Smart" energy management is based on large amounts of data, with a data model, an ICT infrastructure, for the collection of data and its governance, for the integration of data and its sharing, processing and analysis, with security and privacy (Zhou et al., 2016). This management takes into account four main aspects: power generation management, micro-grids and renewable energy management, asset management and collaborative operation, and demand-side management (DSM).

The recommendations and practices to be used in the "smart" network, mainly reliability and low latency, are two goals in which a highly distributed environment allows managing the large amounts of data generated by sensors and meters for application processing. (Jaradat et al., 2015)

The cloud storage solution must be able to store large amounts of heterogeneous data and provide it uniformly. (Fazio et al., 2015). Underlying this solution are a hybrid architecture that combines document and object-oriented strategies to optimize data storage, query and retrieval.

The challenges, from the combination of IoT and big data, bring businesses and technologies that allow cities to realize the vision, principles, and requirements of smart cities applications. Big data can allow you to obtain valuable and valuable information that assist in decision making. (Hashem et al., 2016)

Machine learning (ML-Machine Learning) and artificial intelligence (AI-Artificial Intelligence) can leverage the Internet of Things (IoT) and Big Data (BD) to develop personalized services in smart cities.Chin, Callaghan, & Lam (2017) evaluated, in terms of accuracy, reliability and speed, four ML classification algorithms: Bayes network (BN-Bayes Network), Naïve Bayesian (NB), the J48 decision tree algorithm, and Nearest Neighbor (NN)), on the platform at WEKA ML[34], correlating the effects of weather data (mainly precipitation and temperature) with the short trips made by cyclists in London.

The value of data in large infrastructures enables smart, sustainable and resilient urban planning through master data management (MDM-Master Data Management). MDM systems are based on smart city standards, smart city concept models, "smart" community infrastructure frameworks and web semantics technologies. The aim will be to facilitate the exchange of infrastructure data for smart, sustainable and resilient urban planning. (Ng et al., 2017).

2. Blockchain

In the scope of this work, a systematic review of the literature was carried out, in the investigation of blockchain technology, as a support to the proposal of a trust model

---

[34] WEKA - https://www.cs.waikato.ac.nz/ml/weka/ , accessed on 2019-04-2

applied to smart places, presented at a conference (Brandão et al., 2018b). The review sought answers to the following research questions:

- Q1: What is the evolution over the years in the number of publications on blockchain?
- Q2: What are the main research characteristics analyzed in the blockchain research?
- Q3: What are the application areas of blockchain technology?
- Q4: What are the current research limitations on blockchain research?
- Q5: What are the future research trends and challenges for blockchain?

The review work and the study of adopted ontologies allowed, in summary, to obtain the following answers:

- In the selection of documents from the available bibliographic catalogs, 190 documents were selected for review, which reveal the growing interest in the topic of blockchain technology with the evolution from about 14 documents in 2014 to about 100 already in 2017.
- In these documents, it is also verified that new applications appear in embryonic areas that were not relevant in the primary documents. There is a predominance of the bitcoin system with 38%, the emergence of other cryptocurrencies with 12%, IoT with 28%, the financial sector with 14%, electronic governance with 12%, smart contracts with 10%, smart cities and business, both with 9% and health with 5%.
- The central themes revealed are security issues with around 32%, trust with around 23%, privacy and anonymity with around 18% each and scalability with around 7%.

During the course of the work, the documents for review were updated during the preparation of the thesis, based on their relevance to the theme of the thesis and taking into account the defined ontology.

The blockchain network is revealed as a framework for decentralized data processing, with the characteristics of immutability and self-organization. (W. Wang et al., 2018).

A systematic review of the applications literature (Casino et al., 2018) Blockchain-based technologies identified blockchain technology limitations and future directions to explore in various industries such as supply chain, business, healthcare, IoT, privacy and data management. This emerging technology presents problems and challenges that go through the suitability of the blockchain, the issues of latency and scalability (based on attributes: trust, context, performance and consensus), the challenges of the sustainability of the blockchain protocol, quantum resilience, the adoption of blockchain and interoperability, data management and privacy and security solutions, joint application with big data and artificial intelligence.

Risius & Spohrer (2017) adapt a research framework created to structure current knowledge about blockchain technology, with three groups of activities: design and resources, measurement and value, and management and organization, with four levels of analysis: users and society, intermediaries, platforms, companies and industry. Concluding that the research has been oriented towards technological issues, design and functionalities and not towards governance and value creation.

The smart city uses information and communication technology to integrate and manage physical and social business infrastructures to obtain better services for its citizens, optimizing available resources. The possibility of digital disruption poses important security and privacy challenges. A security framework that integrates blockchain technology with "smart" devices can provide a secure communication platform in a smart city. (Biswas & Muthukkumarasamy, 2016)

Zheng, Xie, Dai, Chen, & Wang (2017) present a comprehensive view on blockchain technology in which the main challenges center on scalability and security issues. They compare some typical consensus algorithms used in different blockchains as per the technical challenges. Several of the concepts and standards presented can help to consolidate these concepts. As ISO TC 307[35] which established the following working groups on: reference architecture, taxonomy and ontology (SG 1); use cases (SG 2); security and privacy (SG 3): identity (SG 4); and "smart" contracts (SG 5). (Anjum et al., 2017)

Blockchain technology has evolved in its characteristics and versions, from version 1.0 blockchain underlying the bitcoin currency to version 2.0 underlying smart contracts 2.0 and, more recently, to 3.0 with the introduction of new architectures and their intended combination. Overcome some of the blockchain's limitations such as the increase in the

---

[35] ISO TC 307 - https://www.iso.org/committee/6266604.html , accessed on 04-8-2019

number of transactions per minute, the possibility of combining various types of records and native exchange with other types of currency. (Brandão et al., 2018b)

2.1. characteristics

This point intends to present the main characteristics of the blockchain that go through the block structure, digital signature and consensus protocols.

The main characteristics of the blockchain are oriented towards: decentralization, in which a third party is not needed to validate, but through consensus algorithms to maintain the consistency of data on the distributed network; persistence, where invalid transactions are detected and once included in the blockchain cannot be deleted or reversed; anonymity, in which the iteration takes place through a generated address, which may not reveal the real identity of the user; and transaction auditing, where they can be easily verified and tracked. (Z. Zheng et al., 2017a)

2.1.1. Block Structure

A generic block consists of a header and a block body. The block header includes:

- The block version, which indicates which set of block validation rules to follow.
- The Merkle tree hash (Merkle tree root hash), which exposes the hash value of all transactions in the block.
- The timestamp, which indicates the current date and time in seconds in universal time since January 1, 1970.
- The nBits field, which indicates the target limit of a valid block hash.
- The Nonce field, which is a 4-byte field, which normally starts with 0 and increases for each hash calculation.
- The tamper-proof data pointer is the hash pointer (Pointer Hash) which indicates a 256-bit hash value that points to the previous block.

The block body is composed of a transaction and transaction (data) counter.

The maximum number of transactions a block can contain depends on the size of the block and the size of each transaction.

The tamper-resistant linked data structure is a block, the tamper-resistant linked-list of blocks is blockchain (as a structural abstraction it will be a list of objects) and the tamper-proof binary tree is the Merkle tree (as a structural abstraction it will be a set of objects). (Gupta, 2017)

2.1.2. Digital signature

Blockchain uses an asymmetric encryption mechanism to validate transaction authentication. Digital signature is based on asymmetric cryptography and is mostly used in untrusted environments. The digital signature is presented as a set of three algorithms: the algorithm to sign it through the sign function (sk, m), the keygen key generation algorithm (n) that results in the private keys sk and public pk and the algorithm verification via the verify function (pk, m, sign (sk, m)). (Gupta, 2017)

Your application will publish the public key pk as your identity and use the secret key sk to prove your identity. So each user has a key pair, private key sk and public key pk. The private key must be reserved and used to sign transactions. Digitally signed transactions are broadcast across the network. Digital signature typically involves two phases: the signing phase and the verification phase.

Sending a message from user X to user Y goes through the signature phase in which user X encrypts his data with his private key and sends the result encrypted with the original data to user Y. In the verification phase, user Y validates the value with the public key of user X, in this way, user Y verifies whether the data has been tampered with or not. The construction of the digital signature commonly used in the blockchain is the ECDSA (Elliptic Curve Digital Signature Algorithm) (D. Johnson et al., 2001).

2.1.3. Consensus Protocols

Consensus in a distributed environment, in the distributed blockchain network, becomes one of the main challenges. In the blockchain, there is no central node that guarantees that the journals (ledgers) on the distributed nodes are the same. The consensus protocol is the core mechanism of a blockchain network that maintains trust in distributed nodes from possible attacks. The following consensus protocols seek to ensure that journals on different nodes are consistent (Z. Zheng et al., 2017a):

- Proof of Work - PoW (Proof of Work)
- Proof of Stake - PoS (Proof of Stake)

- Practical Byzantine Fault Tolerance - PBFT (Practical Byzantine Fault Tolerance)
- Delegated Proof of Stake - dPoS (Delegated Proof of Stake)
- Delegated Byzantine Fault Tolerance - dBFT (Delegated Byzantine Fault Tolerance)
- Proof of Importance - PoI (Proof of Importance)
- *Ripple*
- Stellar Consensus Protocol - SCP (Stellar Consensus Protocol)
- *Tendermint*

The Figure II.10. adapts table II, based on Zheng, Xie, Dai, Chen, & Wang (2017) and table III (Bach et al., 2018), joins them in a single table with the characteristics of the analyzed consensus protocols, through energy savings and the power of tolerance to adversity.

Table II.10 - Characteristics of consensus algorithms

Properties	Algorithms								
	PoW	*PoS*	*PBFT*	*DPoS*	*DBFT*	*POI*	*ripple*	*SCP*	*Tendermint*
Energy saving	No	Partial	Yea	Partial	Yea	Yea	Yea	Yea	Yea
Power of adversity tolerance	< 25% Computing Power	<51% Participation	< 33.3% Replicas	< 51% Validators	< 33.3% Replicas	<50% Importance	<20% Defective nodes	Variable	does not require miners

Adapts table II, based on Zheng, Xie, Dai, Chen, & Wang (2017) and table III (Bach et al., 2018)

PBFT, DBFT, PoI, Ripple, SCP and Tendermint protocols feature energy savings. The consensus protocols, PoS, dPoS and PoI, have greater power of tolerance to adversity (~<51%), although with specific validation mechanisms (participation, validators and importance).

In the following points, we seek to review the characteristics of each of these consensus protocols.

2.1.3.1.   PoW (Proof of Work)

The PoW (Proof of Work) protocol presents a consensus strategy used in the Bitcoin network (Nakamoto, 2008).

In a decentralized network it is necessary to select nodes to record transactions. A simple way is random selection, but random selection can be vulnerable to attacks. So if a node wants to publish transactions in a block, it will have to prove that the node is not capable of attacking the network. This proof translates into significant computations through computation. In the PoW algorithm, each node in the network calculates a hash value from the block header, which contains a nonce, and miners change the nonce frequently to get different hash values. Consensus in PoW requires the calculated value to be equal to or less than a given value. When a node reaches the target value, it will broadcast the block to other nodes and all other nodes must mutually confirm the accuracy of the hash value. If the block is validated,

Miners are the nodes that calculate the hash values. In case of simultaneity in the validation of the blocks, the chain that becomes longer later is considered to be the authentic one. That is, if two nodes create validated blocks simultaneously, miners continue to mine blocks until one branch gets longer and is considered the authentic one.

Several alternative protocols were developed based on PoW, such as the Prime Number Proof-of-Work presented by King (2013) which looks for special strings of prime numbers that can be used across three types of primary strings known as the Cunningham string (Forbes, 1999) of the first type, the second type Cunningham chain and the bigeminal chain (bi-twin) which qualify as proof of work. Main chain is linked to hash lock to preserve security ownership of Bitcoin (Nakamoto, 2008), while a continuous difficulty assessment synthesis is designed to allow the main chain to act as proof of adjustable difficulty.

Other approaches try to optimize by designing Application Specific Integrated Circuits (ASICs) specially designed for PoW computing for blockchain consensus operation, to enable blockchain mining, by taking a significant amount of computation to find a new valid block. (H.Cho, 2018)

2.1.3.2.   PoS (Proof of Stake)

The PoS (proof-of-stake) protocol is an energy-efficient alternative to PoW. Miners in this protocol have to prove ownership of the value. It assumes that those who have more values or more coins are less likely to attack the network. This choice is based on the account balance, as whoever has more values becomes dominant in the network. Other proposals emerge that combine the size of the stake with other criteria to decide which one should shape the next block. Vasin (2014) presents the randomization to predict the next generator through a formula that looks for the smallest hash value combined with the stake size applied to the BlackCoin cryptocurrency. King & Nadal (2012) privileges the choice based on the age of values. The oldest and largest are most likely to mine the next block. PoS saves more energy and is more effective than PoW. Several blockchain networks and platforms initially adopted PoW and moved to PoS as is the case with Ethereum (Wood, 2019) and (Zamfir, 2015)

2.1.3.3. PBFT (Practical Byzantine Fault Tolerance)

The Practical Byzantine Fault Tolerance protocol PBFT (Practical Byzantine Fault Tolerance) is based on a replication algorithm to tolerate Byzantine faults (Castro & Liskov, 1999) where they can allow systems to continue to function correctly even when there are errors, although not all errors are admissible, particularly when they occur on all nodes.

The Hyperledger Project project [36] uses PBFT as its consensus algorithm, as PBFT can handle up to a third of malicious Byzantine replicas. A new block is determined in one rotation. In each rotation, a primary will be selected according to some rules and is responsible for recommending the transaction. The entire process can be divided into three phases: pre-prepared, prepared and committed. In each phase, a node will enter the next phase if it receives votes from more than two-thirds of all nodes. This implies that in the PBFT protocol each node is known by the network.

2.1.3.4. dBFT (Delegated Byzantine Fault Tolerance)

The dBFT (Delegated Byzantine Fault Tolerance), described in the Neo whitepaper[37]. is based on professional nodes that are recommended to record transactions. The dBFT consensus mechanism uses Byzantine fault-tolerant consensus through proxy voting. The bearer of the Neo token[38] can choose the counter it supports and the selected group of counters reach consensus and generate new blocks. Voting on the Neo network continues in real time and not with a fixed deadline. Clients are of two types: counters

---

[36] Hyperledger Project - https://www.hyperledger.org/ , accessed on 2019-04-18
[37] Neo Whitepaper - https://docs.neo.org/en-us/whitepaper.html , accessed on 2019-04-18
[38] We Neo - https://neo-ngd.github.io/reference/How-To-Become-NEO-Consensus-Node.html , accessed on 2019-04-18

and common nodes where common nodes do not participate in consensus, only vote on the counter node they want to support (delegating the vote) and successfully chosen counter nodes are included in the consensus process. If at least two-thirds of the counters agree that the transaction is valid, it remains consolidated on the blockchain and the following consensus rotation will be initiated through another randomly selected counter. (Bach et al., 2018)

2.1.3.5. *dPoS (Delegated Proof of Stake)*

The dPoS (delegated proof of participation) presents as the main difference the type of democracy in which the PoS is a direct democracy, while the dPoS is a representative democracy. Stakeholders elect representatives to generate and validate blocks. As there are far fewer nodes to validate the block, the block can be confirmed faster, leading to faster confirmation of transactions. Network parameters such as block size and block intervals could be adjusted by delegates. Users don't need to worry about rogue delegates as they can be eliminated easily. The BitShares ecosystem[39] uses dPoS to find the most efficient solutions to achieve distributed consensus.

2.1.3.6. *POI (Proof of Importance)*

An account may be eligible for an "Amount Calculation" if it contains at least 10,000 invested XEMs. Given eligibility, Importance is calculated based on the amount of XEM purchased, the account's rank within the network (found using the NCDawareRank algorithm), a weighting factor based on the account's topological location, and two suitable constants determined by the NEM network.[40]

2.1.3.7. *Ripple (Ripple Protocol Consensus Algorithm)*

Ripple (Schwartz et al., 2014) is a consensus algorithm that uses collectively trusted subnets within the larger network. In the network, nodes are divided into two types: server to participate in the consensus process and client to transfer funds only. Each server has a unique node list (UNL). The UNL is important for the server. When determining to place a transaction in the ledger, the server would query the nodes in the UNL and if the contracts received reached 80%, the transaction would be recorded in

---

[39] Bitshares - https://bitshares.org/ , accessed on 2019-04-18
[40] NOR - https://nem.io/ , accessed on 06-18-2019

the ledger. For a node, the ledger will remain correct as long as the percentage of defective nodes in the UNL is less than 20%.

2.1.3.8. SCP (Stellar Consensus Protocol)

The Stellar Consensus Protocol (SCP) (Mazieres, 2015)is a Byzantine protocol agreement that derives from the PBFT. In PBFT, each node has to consult other nodes, while SCP gives participants the right to choose which set of other participants to believe.

2.1.3.9. Tendermint[41]

The Tendermint (Kwon, 2014) is a Byzantine consensus algorithm. A new block is determined in one rotation. A bidder would be selected to transmit an unconfirmed block in this rotation. The process can be divided into three steps: step prediction, in which validators choose whether to transmit a prediction for the proposed block; recommendation step, where if the node has received more than two-thirds of the proposed block, it will transmit a pre-notice for that block and if the node has received more than two-thirds of the pre-commitments, it enters the confirmation step; confirm step, where the node validates the block and transmits an acknowledgment to that block, if the node received two thirds of the acknowledgments, it accepts the block and in contrast to PBFT, the nodes need to lock their coins to become validators and if a validator is considered dishonest, would be punished.

2.1.3.10. Other consensus algorithms

A consensus algorithm must be efficient, secure, and convenient. (Z. Zheng et al., 2017b)

New consensus algorithms are being developed to try to solve specific problems of new Blockchain applications.

The PeerCensus (Decker et al., 2014) Its purpose is to decouple block creation and transaction confirmation so that consensus speed can be increased.

Otherwise, Kraft (2016), to solve the problem of bitcoin security compromise due to high block generation rate, introduces a consensus method to ensure that a block is generated at stable speed. So the string selection rule GHOST (Greedy Heaviest-

---

[41] Tendermint - https://tendermint.com/ , accessed on 06-12-2019

Observed Sub-Tree)(Sompolinsky & Zohar, 2013) is proposed to solve this problem. Instead of the biggest branching scheme, GHOST ponders the branches and miners that could choose the best one next.

Chepurnoy, Larangeira, & Ojiganov (2016) presented another consensus algorithm for peer-to-peer blockchain systems where anyone who provides non-interactive evidence of recoverability, for the past state snapshots, agrees to generate the block. In this protocol, miners only need to store old block headers instead of entire blocks.

The research carried out by Wang et al. (2018), on consensus mechanisms and mining management in blockchain networks, examined blockchain consensus protocols from three perspectives: blockchain network implementers, consensus participants, and blockchain network users. It concludes that in blockchain networks, Byzantine faults cause faulty nodes to exhibit arbitrary behavior, including malicious attacks, node errors, and connection errors.

2.2. Merkle Tree

A Merkle tree, also known as a binary tree hash, is a data structure used to summarize and increase efficiency to verify the integrity of large data sets, allowing for a higher-level view. (Merkle, 1988)

The construction of the Merkle tree involves the use of MD5 hash functions. SHA-3, and SHA-256, which originate unique values, which also reproduce inputs with the current state of a given input. Merkle binary trees project as new information emerges in the data structure and each node has at most two children, which combines two inputs together to obtain a single output. From several pairs of entries (eg log.csv files) the tree structure develops according to theFigure II.2 presented below:

Figure II.2 - Merkel's tree

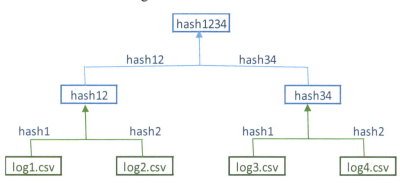

Merkle trees allowcheck if an input data was included in a dataset and in what order. They can also allow compressing large datasets by removing superfluous branches, keeping only those needed for the proof.

The data layer provides the related data blocks using techniques such as asymmetric encryption, timestamp, hash algorithms and Merkle trees and other optimizations. In the blockchain system, each compute node that wins the consensus competition will have the power to create a new block, joining all related data and generated within a specific time period and structured in the Merkle tree. (Yuan & Wang, 2016)

With the use of Merkel's trees, we have the ability to verify if a transaction is included in a block, without having to access the entire chain, and with the possibility of the cryptocurrencies of thesimplified payment verification (SPV - Simplified Payment Verification). Nodes do not maintain the full blockchain and through SPV nodes they use Merkle paths to verify transactions. SPV nodes only use the block header, without knowing the transactions, they just verify the transactions and link that specific chain to the transaction of interest, establishing a link between the transaction and the block that contains it, using a Merkle path. (Kim et al., 2017)

Cho, Park, & Lee (2017) propose blind voting, in which each node is independent, among the nodes, using a block hash and a Merkle tree, where a large number of decisions occur dynamically, without knowing which node to select. Nodes store transferred transactions in a transaction pool according to a certain even number. The Merkle tree is dynamically created each time, so it is a method to eliminate traceability, to block the generator and to collect a large number of decisions.

The main characteristics of Merkel trees are as follows:

- It is verified that if the nodes have the same transaction using the Merkel tree feature, the value of the Merkel tree root will be the same for all nodes. If the transaction is committed the hash value of the Merkle tree root will be incorrect.
- Using Merkle trees allows for greater efficiency, overall performance and scalability.
- Merkle trees become an important component of blockchain technology with scalability and processing power challenges.
- The most common and simplest form of Merkle tree is the binary Merkle tree through the Merkle proof engine. Merkle's proof consists of obtaining the root

hash of the tree and the branches consist of the nodes with the hash that go up along the path from the node to the root. The proof verifies that the hashing is consistent across the tree path.

- All transactions in the block are combined at the root of the Merkle tree. The root of the Merkle tree ensures the integrity of transactions as changing transactions causes a value that is totally different from the value of the root of the Merkle tree. (Lei et al., 2017)

As an application example, the Merkle proofs on the Ethereum platform have in each block header three trees for three types of objects, transactions, receipts and the state, to obtain verifiable answers to various types of queries and questions.

The trees used on Ethereum are more complex using the Merkle-Patricia tree. Merkle binary trees are data structures for preferentially authenticating information in a list format, with a series of blocks in a row, or for transaction trees where no matter how long it takes to edit a tree once created is immutable. State trees become more complex. In Ethereum, the state tree briefly consists of a key-value map, where the keys are the addresses and the values are "statements", cataloging the balance sheets/balances, the nonce, the code and the storage for each account (Storage is itself a tree). The contract status is stored in a balance mapping that associates the users address with a balance. (Wohrer & Zdun, 2018)

The desired data structure must be able to quickly calculate the new tree root after an insert, update, edit or delete operation, without having to recalculate the entire tree through two secondary properties: tree depth limitation preventing service (DoS) and the root of the tree depends only on the data and not on the order of updates.

The Patricia Merkel tree[42] aims to achieve these properties simultaneously. Each node has 16 children and the path on each node is determined by hex encoding, starting at the root, going down to 6, then the fourth, and so on until reaching the end. The storage root is the 256-bit hash of the root node of a Merkle tree that represents the contents of the account. Patricia Merkle trees are used to store all bindings (key, value) on the Ethereum platform. The block header contains the three roots of three representations: status, transactions and receipts.(Vujicic et al., 2018)

---

[42] Patricia Merkel Tree - https://github.com/ethereum/wiki/wiki/Patricia-Tree , accessed on 05-10-2019

2.3. Blockchain Platforms

The different approaches to blockchain technology are associated with different platforms. THETable II.1, adopted from figure 1 of Anjum, Sporny, & Sill (2017), presents and compares the different types of blockchains through the main principles of information security, which include confidentiality, information availability, integrity, non-repudiation, provenance/origin, pseudonymization and selective communication. The colors of the green, yellow and red letters represent the evaluation traffic light system with the respective degree of compliance considered (Green - complies, Yellow - available, Red - does not comply).

Table II.11 - Blockchain information security analysis

Principle	Blockchain security features					
	Bitcoin	Ethereum	stellar	IPFS	Blockstack	hashgraph
Confidentiality	No	No	No	Hash-based content addresses	No	No
Availability of information	mirror block	mirror block	ledger mirroring	Graph and mirror file	Block Mirroring / DHT Mirroring	hashgraph / mirroring; Optional event history
Integrity	Multiple block checks	Multiple block checks	Latest block verification	Hash-based addressing content	Multiple block checks	Consensus with probability a
No repudiation	digital signatures	digital signatures	digital signatures	digital signatures	digital signatures	digital signatures
Provenance	Transaction inputs/outputs	Ethereum state machine and transition functions	Digitally signed ledger transition instructions	Digital signatures and versioning	Transaction inputs and outputs and virtual string references	hashgraph / mirroring; Optional event history
Pseudonymization	the public keys	Public keys and contract addresses	the public keys	the public keys	Public keys, but public information stimulus	Not supported; could be layered
Selective communication	No	No	No	No	Selective access to encrypted storage	Not supported; could be layered

Adapted from figure 1, from Anjum, Sporny, & Sill (2017)

In this table, it can be seen that the platforms under analysis present at least one analysis item in red, which does not comply, of selective communication or confidentiality.

The Table II.12, adopted from figure 2 of Anjum, Sporny, & Sill (2017), presents and compares the main performance characteristics of the various types of blockchains, through the main characteristics associated with blockchain technology, which include consistency, system availability, fault tolerance, scalability, latency, auditability, resilience, resistance to denial of service (DoS) and system complexity. The colors of the green, yellow and red letters represent the evaluation traffic light system with the respective degree of compliance considered (green - complies, yellow - available, red - does not comply). These types of blockchain and respective platforms are not limited to the types presented,

Table II.12 - Analysis of blockchain performance characteristics

Principle	Blockchain performance characteristics					
	Bitcoin	Ethereum	stellar	IPFS	Blockstack	*hash*graph
Consistency	Block Checks. 30 to 60 minutes	Block Checks. 20 to 60 minutes	Single Block Scan Less than 1 minute	P2P mirroring. Limited primarily by network I/O. Several seconds for files less than 128 KB.	Block Checks. 30 to 60 minutes	Consensus with probability one; Byzantine agreement, but attackers must control less than a third
System availability	Block Checks. 30 to 60 minutes	Block Checks. 20 to 60 minutes	Single Block Scan Less than 1 minute	Single storage request response. Several seconds for files less than 128 KB.	Block Checks. 30 to 60 minutes	Virtual voting; Tough DoS no proof of work, quick talk
Fault tolerance	Longest chain wins	Longest chain wins	Last voting block always has consensus.	Hash content address. Highly resilient against network partitioning.	Longest chain wins	Virtual voting. Tough DoS no proof of work, quick gossip
Scalability	Block size. 7 transactions per second	Block size. 7-20 transactions per second	Thousands to tens of thousands of transactions per second.	Thousands to tens of thousands of transactions per second. Scales linearly as nodes are added.	Block size. 7 transactions per second	Thousands to tens of thousands of transactions per second. Limited only by bandwidth
Latency	Block Checks. 30 to 60 minutes	Block Checks. 20 to 60 minutes	Single Block Scan Less than 1 minute.	Single storage request response. Several seconds for files less than 128 KB.	Block Checks. 30 to 60 minutes	Virtual voting; limited only by gossip protocol, exponentially fast
Auditability	Total	Total	Total	Difficulty	Total	configurable
Resilience	Total	Total	Total	Fails if nodes storing data fail	Total	Total
Resistance to denial of service	Spend Bitcoins	spend ether	spend Stellar	Files are only mirrored if requested	Spend Bitcoins	Signed Status / Proof of Participation / <1/3 attackers
System Complexity	Average	High	Average	Average	Average	Low but system wide

Adapted from figure 2, from Anjum, Sporny, & Sill (2017)

In the following sub-points, some of the main analyzed platforms will be detailed, Hyperledger Fabric, Stellar, Ethereum and InterPlanetary File System that will be able to support the types of blockchain, Ethereum, Stellar and IPFS. However, the platforms and types are not limited to the platforms analyzed below and in Annex I - Comparative table of blockchain platforms, which presents a list of platforms with their main characteristics.

2.3.1. Hyperledger Fabric (HLF)

The Hyperledger project[43]it was based on several requirements that include: private transactions and confidential contracts to ensure confidentiality and privacy; identity and auditability to support identity exchange and auditing, through the possibility of identity anonymity; interoperability, through independent components, through interaction when information is exchanged and used by these components; portability in heterogeneous computing environments, by abstracting from component interfaces, separating them from the environments; the architecture based on four categories, which are made available as identity, policy, blockchain and smart contract services.

The HLF presents a consensus that is reached when the order and results of a block's operations have met the explicit criteria, which go through: the query and accounting update, using key-based lookups, range queries, and composite key queries. ; read-only history queries; operations that contain the signatures of all the peers to be endorsed and that are submitted to the ordering service, through the validation of transactions that comply with the policies of, in particular, contribution; the accounting of a channel that contains a configuration block with the definition of policies, access control lists, and other pertinent information; and by the channel that allows encryption of different certificate authorities.

The main functions include: permissive blockchains with immediate purpose; environments for the execution of "smart" contracts; PBFT-based consensus modules (Castro & Liskov, 2002), in No-op (consensus ignored) to handle in standard mode while the code is in pre-release version, to debug (debug) and the SIEVE prototype (to perform operations, compare output in copies and look for disparities) as modular solution to replicate non-deterministic applications in a BFT system (Cachin et al., 2016); an event framework that supports predefined and custom events; the SDK client (Node.js) to interact with the blockchain network; and support for basic REST (Representational State Transfer) APIs and CLIs (comand-line interfaces) (Cachin, 2016).

The investigation carried out by Sousa, Bessani, & Vukolić (2017) on the HLF (Hyperledger Fabric) platform allows extensibility and support for various ordering

---

[43] Hyperledger Fabric - https://github.com/hyperledger/fabric , accessed on 03-10-2019

services for building the blockchain. The implementation and evaluation of the BFT (Byzantine Fault-Tolerant) parcel service for HLF allowed it to reach up to ten thousand transactions per second and irrevocably write a transaction on the blockchain in half a second.

2.3.2. Stellar

The platform uses the Stella consensus mechanism. The Stellar Project[44] is an open-source project, with a distributed payments infrastructure, which connects banks, payment systems and people. The platform stellar allows the construction of mobile wallets, banking tools, and "smart" devices and provides HTTP API RESTful servers called Horizon, which connect Stellar Core to the Stellar network.

A lumen (XLM) is the native asset of the Stellar network, a unit of digital currency such as a bitcoin.

2.3.3. Ethereum

The Ethereum platform[45] is decentralized, capable of executing smart contracts and decentralized applications supported by blockchain technology.

The main features include: applications that work according to the way they were programmed, through the contract that is immutable, without censorship, fraud and third-party interference; has a decentralized virtual machine, the Ethereum Virtual Machine (EVM - Ethereum Virtual Machine), to run scripts on public nodes; and with the Casper protocol(Baliga, 2017) the Ethereum network went from Proof-of-Work (PoW) to Proof-of-Stake (PoS).

The Casper Protocol[46], released on the Serenity version of the Ethereum platform, uses the concept of deposits and security bets to gain consensus. In this protocol, the nodes connected to the Ethereum system make significant security deposits defined by the protocol, becoming connected validator nodes that demonstrate commitment and

---

[44] Stellar - https://www.stellar.org , accessed on 11-03-2019

[45] Ethereum - https://www.ethereum.org/ , accessed on 2019-02-20

[46] Ethereum Casper Protocol - https://blockgeeks.com/guides/ethereum-casper/, accessed on 2019-07-20

interest in advancing in the Ethereum blockchain network, betting their security deposits.

This initial list of validators is linked through the Casper contract, which can evolve based on new nodes and older nodes leave the system. Each validator is pseudorandomly selected to produce a block from the active validator set, with the probability of selection linearly weighted by the deposit of each validator. If a validator is offline, a different validator will be selected and this process will be repeated until an online validator is found and creates a block. If a validator produces a block that is included in the chain, it receives a block reward equal to the total Ether in the active validator's pool. If the validator produces a block that is not included in the chain, the protocol works in such a way that the validator loses the security deposit equal to the block reward.(Baliga, 2017).

The Ethereum platform allows: to facilitate the security of crypto-assets, the writing, implementation and use of smart contracts; cryptocurrencies are created; democratic autonomous organizations (DAOs - Decentralized Autonomous Organization) are created; and if you use development tools in Go, C ++, Python, Java, etc. An Ether (ETH) is the native asset of the Ethereum network, a unit of digital currency.

2.3.4. InterPlanetary File System (IPFS)

The InterPlanetary FileSystem[47] (IPFS) (Benet, 2014)is a platform to support a peer-to-peer distributed file system that seeks to interconnect devices to the same file system. IPFS can be seen as a single BitTorrent location, which exchanges objects within a Git repository. Your organization goes through a generalized Merkle DAG (Directed Acyclic Graph) (data structure similar to a Merkle tree, less rigorous since the DAG may not be balanced and its nodes may contain data), on which it is possible to build systems of versioned files, blockchains and even a permanent web. IPFS combines a distributed hash table, block swapping and a self-certified namespace, tends to have no single points of failure (with new versions, performance improvements, and node growth) and nodes don't need to trust each other.

---

[47] IPFS - https://ipfs.io/ , accessed on 2019-04-10

In the technical specifications of the IPFS protocols[48] The IPFS protocol, the network layer (libp2p covering the network and routing), the registers, the naming and recording systems (IPRS-InterPlanetary Record System and IPNS - InterPlanetary Naming System), the data structures and formats (IPLD-InterPlanetary Linked Data, unixfs and multiformats), the mutable files/filesystem (Virtual File System interface, on linux and on top of MerkleDAG), the exchange of blocks (bitswap inspired by BitTorrent), specific internal components (service blocks and blocking, DAG and DAG service, data import, IPFS node local repository specification), public APIs (Core Core API, HTTP API and CLI), key management (KeyStore, KeyChain) and the Protocol Driven Development (PDD-Protocol Driven Development).

In summary, the main components of IPFS are: the distributed hash table whose nodes can store and share data without central coordination; the IPNS (InterPlanetary Naming System) which allows exchanged data to be instantly pre-authenticated and verified using public-key cryptography; the Merkle DAG which allows data to be uniquely identified, tamper-proof and permanently stored; and allows accessing previous versions of edited data through the Version Control System (Git). (Benet, 2014)

Zheng, Li, Chen, & Dong (2018) propose a blockchain data storage model that shows how to use IPFS networks to reduce blockchain data storage, through a proof of concept for miners to be able to store less blockchain data in real scenario and how new nodes can quickly sync with the network . As a result, they calculated the compression rate against bitcoin (can reach 0.0817) and verified better characteristics of the scheme, in terms of storage space, security and node synchronization speed.

2.4.   Blockchain comparison with other technologies in IoT environment

Hankerson, Vanstone, & Menezes (2003) reveal the five fundamental goals for secure communications: confidentiality, to keep data secret from everyone except those authorized to see it; data integrity, to ensure that data has not been altered by unauthorized means; data source authentication, corroborate the data source; entity authentication, to corroborate the identity of an entity; and non-repudiation, to prevent an entity from denying prior commitments or actions.

---

[48] SPEC IPFS - https://github.com/ipfs/specs , accessed on 2019-03-10

The IoT layers, for smart solutions, are represented in the Table II.13, adapted from Yang, Wu, Yin, Li, & Zhao (2017), with existing solutions.

Table II.13 - IoT layered analytics

layers	Solutions
application layer	*Smart places, smart city*, smart home, healthcare system, energy management, environmental monitoring, industrial internet, connected vehicles
transport layer	*Transport Layer Security* (TLS), Datagram Transport Layer Security (DTLS)
Network layer	*Low power Wireless Personal Area Networks* (6LoWPAN), 6LoWPAN/IPSec, IPSec, IPSec's Authentication Header (AH), Encapsulation Security Payload (ESP), End-to-End (E2E), IEEE 802.15.4, IPv6
Layer of perception	*Wireless Sensor Networks* (WSNs), Implantable Medical Devices (IMDs), Implantable Cardioverter Defibrillator (ICD), Radio-Frequency Identification (RFID), Global Positioning System (GPS).

Adapted from Yang, Wu, Yin, Li, & Zhao (2017)

The Table II.14, adapted from table 3, fromMohamad Noor & Hassan (2019) seeks to summarize the main IoT security challenges and possible mitigation measures.

Table II.14 - Current challenges in IoT security and mitigation measures

Layer	Security Challenges	Mitigation
Perception	Abnormal sensor node detectionThe encryption algorithms of choice and the key management mechanism to use	Fault detection algorithm, decentralized intrusion detection system
		Public key encryption due to large-scale network
		Slot reservation protocol
		Access control, DoS attack mitigation
Network application	Sender Data and Anonymity Device Vulnerabilities Enabling IPSec Communication with IPv6 Nodes Embedded and Configurable computer systems.	Research on the suitability of IPv6 and IPSec for secure communication.

Adapted from table 3, from Mohamad Noor & Hassan (2019)

Public key cryptography solutions imply adequate key management that includes generation, distribution, storage, updating and destruction of secret keys. Low-power public-key cryptography algorithms applied in IoT and for wireless sensor networks go through algorithms such as: Rabin's Scheme[49], the NtruEncrypt[50] and Elliptic Curve Cryptosystems (ECC) (Gaubatz et al., 2005). The ECC defines the set of different asymmetric cryptographic key exchange and agreement protocols such as ECDH (Elliptic-curve Diffie-Hellman), ECDSA (Elliptic Curve Digital Signature Algorithm) and ECMV (Elliptic Curve Menezes Vanstone). Key distribution can be divided into four groups: broadcast key distribution, group key distribution, master key pre-distribution, and paired key distribution.

---

[49] Rabin, point 8.3 - http://cacr.uwaterloo.ca/hac/about/chap8.pdf , accessed on 06-10-2019

[50] Ntru Encrypt - https://www.onboardsecurity.com/products/ntru-crypto/ntru-resources , accessed on 06-10-2019

Suarez-Albela, Fernandez-Carames, Fraga-Lamas, & Castedo (2018) compared the performance of the ECDSA and RSA cipher suites (Rivest et al., 1978), and evaluated the impact of different ECC curves and RSA key sizes using resource-constrained IoT nodes. In the selected scenarios, ECDSA(D. Johnson et al., 2001) is a better alternative than RSA for securing IoT devices with limited resources.

The security problems of IoT solutions can affect the deployment and adoption, implying new research in security in IoT technologies, to try to solve the challenges and risks, in detecting weaknesses to minimize vulnerabilities, through reliable technologies and the possibility of integration and of security mechanisms. The IoT security areas, which began by adapting existing protocols and algorithms to the IoT context, include: forensic analysis, security engineering, intrusion detection, resilience, privacy, trust in the social IoT, trust, self-healing, security policies, architectures, access control, authentication, identity, communications, operating systems, encryption, non-clonable physical functions (PUF-Physical Unclonable Functions), physical layer security (PLS) and hardware security. DLT (Distributed Ledger Technology) technologies, in this case the blockchain, provide reliable and decentralized operations, namely exchange of tokens, storage of metadata and execution of programs and other services, which, when used in IoT, can track physical elements. and digital, creation of IoT metadata, access control decentralized management services, decentralized and reliable firmware update. (Roman-Castro et al., 2018)

Both types of new technologies go through SDN (Software Defined Network) and blockchain that converge with IoT security solutions. SDN separates network control and data control, enabling centralized control and dynamic network management for allocating resources to IoT devices. the work ofShaghaghi, Kaafar, Buyya, & Jha (2018)establishes a set of requirements for an IoT security solution and analyzes existing solutions against these requirements. Blockchain can solve the problems of reliability, security, scalability and QoS (Quality of Service). Blockchain applied to IoT allows for decentralization, pseudo-anonymity and secure transactions. (Mohamad Noor & Hassan, 2019)

Blockchain technology applied to large-scale IoT systems allows: data to be tampered with; trust less although establishing the possibility of message exchange, more robust and more reliable; data is more private; record historical actions and past transaction data of "smart" devices; distributed files are shared; the single supervisory authority is eliminated; costs in infrastructure development are reduced; confidence is increased; and speed up transactions. (Kumar & Mallick, 2018)

The essential activity of a blockchain network is to ensure that the networks trusted nodes agree on a single, tamper-proof transaction record. The network must tolerate that part of the nodes that deviate from that canonical register. The blockchain network can be summarized in four levels of implementation: the data organization and network protocols, the distributed consensus protocols, the autonomous organization framework based on distributed virtual machines (VM) and the implementation of applications with human-machine interfaces. (W. Wang et al., 2018)

In summary, the issues and challenges that go through the use of blockchain technologies, go through more reliable and convenient services, but in which security issues remain in the need to be cautious in choosing the solutions to adopt. (Lin & Liao, 2017)

Blockchain technology can provide better security, especially for sensitive data and in blockchain applications where transparency and immutability need to be promoted. (Stephen & Alex, 2018)

2.5. Architectures and Taxonomy

The taxonomy proposed by Xu et al. (2017) reveals the key architectural features of blockchains and the impact on their key design decisions. The design decision as a blockchain project is characterized by the data structure, the consensus protocol, the protocol configuration and the new blockchain, oriented towards thoughtful security and prudent scalability. These options are rated in detail by their impact through key properties such as cost efficiency, performance and flexibility.

The taxonomy can show the impact of different design choices on quality attributes and allows quality attribute analysis to provide a basis for comparison.

The design of new forms of distributed software architectures involves defining where agreement and sharing can be established without relying on a central integration point. Blockchain is considered as a piece of software linking with a complex internal structure, of several configurations and different variables, which implies defining a taxonomy to classify, compare and evaluate the impact on software architectures.

The Figure II.3, adapted from figure 1 of Xu et al. (2017), presents the design process for blockchain-based systems and features the following four design steps:

1. Architectural design regarding decentralization, where decentralization focuses on the responsibility and capability of a central location or authority, where there is a range of possibilities between decentralization and centralization, with the two options for partial decentralization: permission and verification.
    a. Permission, rather than anonymous public participation, on a blockchain may be allowed one or more authorities to act as an entry to participation.
    b. Verification, in which the execution environment of a blockchain is self-sufficient and can only access information present in a transaction or in the blockchain's transaction history, and the states of external systems are not directly accessible.
2. The architectural design in relation to storage and computing, in which blockchains provide unique properties, in terms of the amount of computing power and data storage space, which are still limited. Cost efficiency, performance and flexibility are key design decisions for using blockchain, which includes choosing what data and computation should be put on-chain and what should be kept off-chain.
    a. Item data addresses a common procedure for data management in blockchain-based systems to store raw data off-chain, and to on-chain only metadata, critical small data, and hashing the raw data.
    b. Item collection addresses the concept of data collection common in blockchains when using blockchains as a ledger.
3. Architectural design in relation to blockchain configuration, where various configuration options for using blockchain are considered, as a design decision it is intended to have an objective to use a public blockchain, consortium/community blockchain or private blockchain.
4. In the other architectural and implantation projects, the other design options involve anonymity and incentives, and the implantation impact.
    a. Anonymity refers to the different techniques that have been proposed to preserve anonymity on the blockchain.

b. The incentive allows blockchains and their applications, particularly public blockchains, to introduce financial incentives or reputation and ranking mechanisms, to get miners to join the network, to validate transactions and correctly generate blocks.

c. Blockchain deployment also has an impact on the quality attributes of the system.

Figure II.3 - Design process for blockchain supported systems

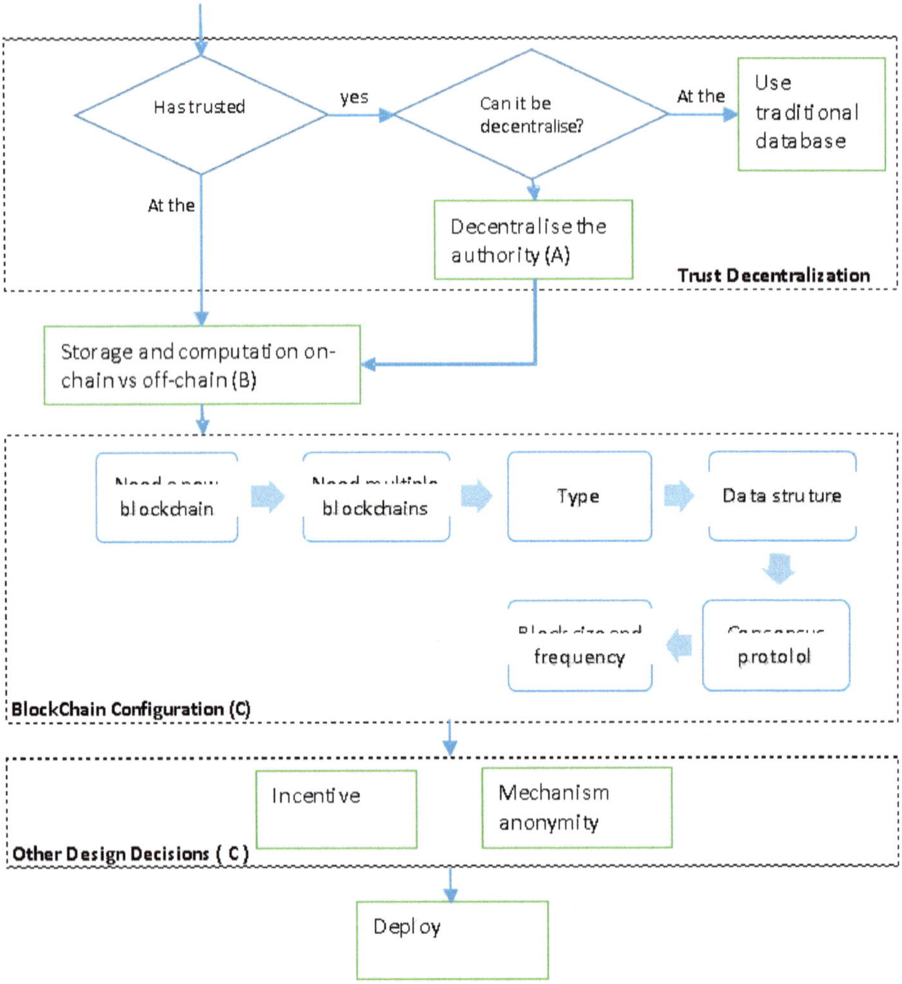

Adapted from figure 1, from Xu et al. (2017)

Blockchain technology is used in scenarios where no single trusted authority is required and the trusted authority may be decentralized or partially decentralized. Blockchain projects related to design decisions regarding (de)centralization are weighted with their impact (1 - least favourable, 2 - neutral and 3 - most favourable) on quality properties (fundamental properties, cost efficiency, performance and flexibility) are discussed in the Table II.5, adapted from table I of Xu et al. (2017). Blockchain limitations are the next decision about dividing computing and storing data between on and off chains.

Table II.15 - Blockchain project related design decisions

Decision	Option	Impact			
		Fundamental properties	Cost efficiency	Performance	Failure Points
fully centralized	Single-vendor services (eg governments, courts)	1	3	3	1
	Services with alternative providers (e.g. banking, online payments, cloud services)				
Partially centralized and partially decentralized	Blockchain with permission with permissions for fine-grained operations on the transaction level (for example, the permission to create assets)	two	two	two	-
	Permissioned blockchain with miners with permission (write) but normal nodes with lower permission (read)				
fully decentralized	Blockchain with less permission	3	1	1	Majority (us, power, participation)
verifier	Single reliable verifier of network signals (external verifier valid transactions; internal verifier uses previously injected external state)	two	two	two	1
	M-of-N network trust checker	3	1	1	M
	Ad hoc trust checker by involved participants	1	3	two	1 (choice per ad hoc)

Adapted from table I of Xu et al. (2017)

Design decisions related to blockchain project design decisions regarding storage and computation are analyzed through their relative impact (1 - not very favorable, 2 - least favorable, 3 - more favorable and 4 - very favorable) on the properties. of quality (fundamental properties, cost efficiency, performance and flexibility) and are discussed in Table II.17, adapted from Table II of Xu et al. (2017). After that, a set of design decisions about the need for blockchain configuration has to be made, such as the blockchain type, consensus protocol, block size and frequency.

Table II.16 - Blockchain project related to design decisions.

Decision		Option	Impact			
			fundamental properties	cost efficiency	performance	Flexibility
item data	In jail	Embedded in the transaction (Bitcoin)	4	1	1	two
		Embedded in the operation (Public Ethereum)		4	1	3
		Variable "smart" contract (Public Ethereum)		two	3	1
		Event log "smart" contract (Public Ethereum)		3	two	two
	off chain	Private/third-party cloud	1	^ Insignificant Kb	4	4
		Peer-to-Peer System		4	3	3
Item collection	In jail	smart contract	4	4 (public)	4	1
		separate chain		1 (public)	1	4
computing	In jail	transaction restrictions	4	1	1	1
		smart contract				
	off chain	Private/third-party cloud	1	4	4	4

Adapted from table II of Xu et al. (2017)

Design decisions about blockchain configuration are discussed in Table II.17, adapted from table III of Xu et al. (2017), with the qualitative evaluation (1 - least favorable, 2 - neutral and 3 - more favorable) of the quality properties.

Table II.17 - Blockchain projects related to project decisions about blockchain configuration.

Decision		Option	Impact			
			fundamental properties	cost efficiency	performance	Flexibility
Blockchain Scope		*Blockchain* public	3	1	1	1
		Blockchain community/consortium	two	two	two	two
		Blockchain private	1	3	3	3
Data structure		*Blockchain*	3	1	1	1
		GHOST	two	two	two	1
		BlockDAG	1	3	3	3
		segregated witness	3	two	1	1
consensus protocol	Security approach	proof of work	3	1	1	1
		Proof of recoverability	3	1	1	1
		trial of game	two	two	two	3
		BFT (Byzantine Fault Tolerance)	1	3	3	1
	Scalability Approach	Bitcoin-NG	3	1	1	1
		Off-chain operation protocol	1	3	two	3
		mini-blockchain	two	two	1	two
Protocol settings	Security approach	X-block confirmation	1	1	1	3
		Checkpointing	3	3	3	1
	Scalability Approach	Original block size and frequency	3	n/a	1	n/a
		Mining time increase/decrease block size	1	n/a	3	n/a
New blockchain	Security approach	Embedded mining	3	two	1	1
		Blockchain transaction-level hook popular	two	1	two	3
		Proof-of-burn proof	1	1	3	two
	Scalability Approach	side chains	3	1	1	1
		Multiple private blockchains	1	3	3	3

Adapted from table III of Xu et al. (2017)

The taxonomy can be used to compare blockchains, correctly define the project and allow the evaluation of software architectures supported by blockchain technology, through the characteristics of blockchains and their impact for different decision scenarios based on performance and quality attributes, namely availability, security and performance.

2.6. Blockchain technology applications

Blockchain applications begin to be global and mainly in the domains of smart applications. Next, we review several articles, some of which have already been mentioned, that analyze blockchain technology application projects, mainly in the field of smart places supported by IoT.

Dorri, Kanhere, Jurdak, & Gauravaram (2017) describe the main components of a smart home (IoT devices, local storage, the miner and the local BC) and analyze the various transactions and procedures associated with it, with concerns about security and privacy. The smart home simulation demonstrates that the overall costs incurred by the method described are low and manageable for low-resource IoT devices and are acceptable given the security and privacy benefits offered.

Ra & Lee (2018) present a blockchain-based solution that promotes confidentiality through a group key and through chain-based key management, in a smart home environment, using the Extended Merkle tree and KSI (Keyless Signature Infrastructure) based authentication and communication.

Yli-Huumo, Ko, Choi, Park, & Smolander (2016)present as examples the prototypes of applications developed and suggested for the use of blockchain in other environments, such as IoT, "smart" contracts, smart property, digital content distribution, botnet and P2P transmission protocols, used in an environment decentralized. This shows that blockchain technology is not limited to applications in cryptography.

Hammi, Hammi, Bellot, & Serhrouchni (2018) present a solution in blockchain technology to define virtual safe zones where IoT equipment can identify and trust each other.

Jabbari & Kaminsky (2018) argue that blockchain has the potential to advance supply chains, but further research is still needed and that they identify four categories of questions: How can physical products be linked to the digital ledger? How can blockchain-enabled networks be linked to other external markets? How can the blockchain structure be improved to allow for more complex supply chain structures? How can enough space be reserved to store the amount of information required by supply chains? These issues can be reviewed in the context of data markets supported by blockchain and with the use of smart contracts to interconnect the various supply platforms.

Qu, Tao, Zhang, Hong, & Yuan (2018) present the forms of blockchain structures (BCS-Blockchain Structures) designed to establish the relationship between blockchain and IoT.

IoT devices are analyzed from four perspectives: exploring the most important limitations of IoT devices and what the solutions are; the classification of IoT attacks;

authentication and access control mechanisms and architectures; and security issues in the different layers. (Yang et al., 2017)

Gaetani et al. (2016) present a blockchain-based database to ensure data integrity in cloud computing environments where threats to data integrity and data tampering, which can maliciously affect critical decisions, are intended to be addressed. In cloud computing environments, data owners cannot control important aspects, storage and access control. The use of blockchain to minimize data integrity threats, although it seems to be a natural choice, still has several limitations such as low throughput, high latency and some instability.

Blockchain technology can be used to improve power grid robustness and security (Liang et al., 2018) through a distributed blockchain-based framework for data protection and to increase the self-defense capability of electrical power systems against cyber-attacks. Blockchain features in the proposed framework go through a private network, transaction initiator must be completely automatic, transaction content must be measurement collection, transactions are independent and unrelated, verification of historical blocks before the transaction process voting is unnecessary, the connection speed of the chain is much faster than the classic 7 transactions per second (bitcoin), there is no reward for the node, there is no double spend attack, and the 51% attack is difficult given that the limit is adjustable. The case study was based on the IEEE-118 benchmarking system, which is composed of 54 generators, 118 nodes and 186 branches, used as a basis for cyber attack scenarios between the existing method and the proposed structure, represented in two scenarios. The work shows that blockchain can be considered a promising solution in data security in energy systems.

3. Data Market

At this point, the application of blockchain technology to data marketplaces stands out. This highlight reflects the importance of these data markets in the context of smart cities, especially open data and the need to provide data from reliable suppliers, from reliable sources and with reliable data.

Data markets, in the context of smart places, with the huge amounts of data, mainly from IoT, reveal several centralized and decentralized approaches. in the article (Brandão et al., 2019) developed within the scope of this research work, the decentralized data markets supported by blockchain technology are analyzed as a way of guaranteeing trust in the data supply chain, the actors that intervene in the market and the data sources, through contractualization through smart contracts, the provision of data by the data producers, the control of data flows and access to data.

Data markets can be based on a centralized or decentralized architecture and on three openness models: open, closed or mixed.

Data is based on dynamic data, static datasets, through dynamic and static streams. Access to data is carried out through external or internal APIs, download or blockchain platforms, with the aim of obtaining quality data and information, using user feedback, data integrity and data provenance.

The most important features for participants in data markets are: a data catalog with search functions; access control; tools to create suitable agreements; appropriate monetizations for transactions; monitoring compliance with SLAs and resolving disputes; and tools to assess the quality and reliability of data and its suppliers.

Trust is central to assessing the quality and reliability of data and its providers. Security, with the management, authentication and authorization of identities, is integrated into market solutions.

The role of data markets is crucial in the information and knowledge economy, which makes reliable data fundamental for the automation of rules in machine-to-machine interaction, in machine learning, in the application of AI (Artificial Intelligence) algorithms, in decision-making and to create new business opportunities.

The value of the data will be conditioned by weaknesses and possible cyber breaches, which increase financial losses, involving insurance policies on the data, for commercial protection. Data marketplaces will promote cross-sourcing with various data producers, involving different data formats, namely IoT data and crowdsourcing.

Control of the data supply chain starts at the trusted source, continues in the process of processing and creating Value and ends with the satisfaction of delivery. To ensure security and reliability, four application levels are proposed, adding specific features, supported by blockchain technology.

The problem of trusting the data source can be solved by a blockchain-supported model and architecture that address the four levels of this problem:

- The provision of data;

- The delivery of data;
- Management of access rights and permissions;
- The data producer.

This model intends to guarantee the origin of the base data and the trust in the whole process. The Figure II.4, adapted from figure 3, from (Brandão et al., 2018a) presents the actors, data flows, platforms and forms of blockchain application.

Figure II.4 - Blockchain based data market

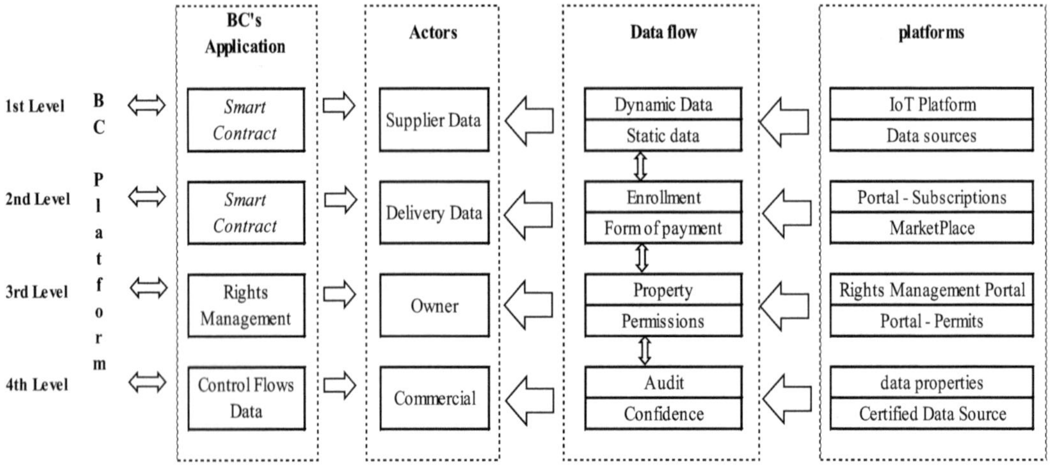

Level	BC's Application	Actors	Data flow	platforms
1st Level	Smart Contract	Supplier Data	Dynamic Data / Static data	IoT Platform / Data sources
2nd Level	Smart Contract	Delivery Data	Enrollment / Form of payment	Portal - Subscriptions / MarketPlace
3rd Level	Rights Management	Owner	Property / Permissions	Rights Management Portal / Portal - Permits
4th Level	Control Flows Data	Commercial	Audit / Confidence	data properties / Certified Data Source

(BC ⇔ Platform form)

Adapted from figure 3, from Brandão et al. (2019)

Ivanschitz, Lampoltshammer, Mireles, Revenko, & Schlarb (2018) present a data market based on multiple decentralized data repositories, allowing access through a central portal, supported by standardized and semantically improved, reliable and consistent metadata, to efficient search and recommendation functionalities supported by a central catalogue. In these decentralized repositories each participating node implements the defined set of services and interfaces such as a data tracker, a metadata mapper, a blockchain node and data management and storage components, and a common conceptual model to enable standard interfaces that facilitate the interoperability and the use of datasets.

The sterling job (Hynes et al., 2018) presents a decentralized data market of private data, with the distribution and preservation of data privacy through smart contracts supported on blockchain. The immutable and irrevocable smart contracts from data providers to consumers represent the interests of their creators by automatically evaluating the data, through mechanisms so that data providers can control the use of their data, through automatic verification of the data consumption contracts and express restrictions, such as pricing and privacy differential. This data market allows the resulting economy to confirm that the interests of all parties are aligned.

In the data market presented by Agarwal, Dahleh, & Sarkar (2018) is considered an algorithmic solution for creating the data market, with a robust real-time matching engine to efficiently buy and sell test data for machine learning tasks. This solution considers the monetization of data and pre-test models to price test data between buyers and sellers. These data are freely replicable, their value dependent on correlation with other data, prediction tasks, accuracy and usefulness of the data.

Chakrabarti & Chaudhuri (2017) discuss how blockchain technology can be used in the data market associated with business processes in the retail sector to benefit customers and retailers, with transparency about the origin of products, combating counterfeiting, more efficient management of the supply chain, supply and improve loyalty management by improving customer profiles.

The proposal to use the data market service model to solve the sharing of scientific data is presented by Ghost (2018), as a platform for sharing data for the scientific community, analyzing the motivational challenges and practices for the exchange of scientific data. Challenges in data sharing include motivating collaboration, multiple data structures and formats, avoiding data incompleteness, duplication and inconsistency, different contexts, different data organization, data dynamics, data ownership management, the creation of a community of data producers and consumers forming a data ecosystem created to guarantee data quality and with adequate prices. The Figure II.5, adapted from figure 1 of Ghost (2018), presents a possible general model for a data marketplace.

Figure II.5 - General data marketplace model

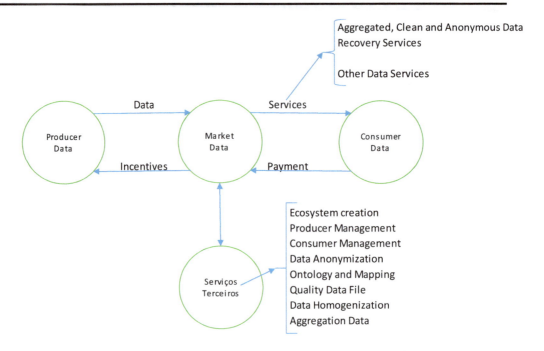

Adapted from figure 1, from Ghost (2018)

The data supply chain is presented in simplified form in Figure II.6, adapted from figure 1 of (Brandão et al., 2019), where suppliers deliver raw data from producers (data generation), who convert data into products for data "warehouses" (data transformation), who store data in distribution centers by providing retailers with processed data (data processing) that filter, aggregate and format them, allowing various visualizations and customizations to deliver them to the user (analytics / use).

Figure II.6 - Simplified data marketplace model

Adapted from figure 1 of Brandão, Mamede, & Gonçalves (2019)

The Figure II.7, adapted from Figure 2 by Brandão, Mamede, & Gonçalves (2019), presents an associated data market model, with the actors of this market: the data owner, the data broker, the Value Added (VA) service provider) and the end user.

Figure II.7 - Data market model and actors

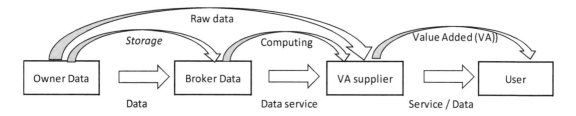

Adapted from figure 2 of Brandão, Mamede, & Gonçalves (2019)

Dao, Alistah, Musat, & Zhang (2018) present the DataBrigh project in the scope of machine learning for decentralized global data exchange, property management and reliable computing. This project seeks to answer the question: how to promote the global exchange of data, where everyone can contribute computing and data to train machine learning applications? The DataBright system is a data market and a reliable computing market, which transforms the creation of training examples and computing sharing into an investment mechanism, in which those who contribute become a shareholder in the data set they created.

Draskovic & Saleh (2017) present a data marketplace, Datapace, for IoT sensor data. This data marketplace is decentralized and blockchain-based. They consider that there are data whose use is to be consumed immediately, otherwise they lose value, and lasting data. But whoever collects the data wants to make an additional profit from that data, beyond the underlying purpose of the data. The Datapace system enables value tokenization and token saving, ensures data integrity through data hash custody, enables smart contract features, and provides network security through PBFT's consensus and immutability features.

Smith, Ofe, & Sandberg (2016) explore the value proposition of an open data market, as an innovation in digital services, seeking to overcome the barriers to adopting open data, through an exploratory case study, considering the five most valued perceived values that go through: the lowest task complexity, greater access to knowledge, greater possibilities to influence, less risk and greater visibility. In this work, it was found that for users of open data, Valor is in the central portal that provides the best access to open data and associated support services, and to be able to access open data providers in knowledge sharing activities, allowing the knowledge transfer within ecosystems.

The Enigma platform (Zyskind et al., 2015) decentralized computing is intended to ensure the privacy of different parties that can store data together and perform calculations on the data while simultaneously keeping the data private. Enigma's computational model is based on an optimized version of secure multiparty computation, guaranteed by a verifiable secret sharing regime. Enigma has an off-chain distributed hash table (DHT-Distributed Hash-Table) that is accessible through the blockchain, which stores only the data references. Private data must be encrypted before storage and access control are programmed into the blockchain.

The MARSA data market (Cao et al., 2016) features a dynamic cloud-based marketplace of near real-time human sensing data so different stakeholders can sell and buy this type of data. This project presents techniques for selecting what types of data and how to manage data contracts based on different cost models, data quality, and data rights. The platform considers solutions for different data transfers to enable an open and scalable communication mechanism between sellers (data providers) and buyers (data consumers).

The new Common Vehicle Information Model (CVIM) (Pillmann et al., 2017) can leverage the creation of big data markets with vehicle data. The European AutoMat project aims to develop an open market by providing a single point of access to vehicle data regardless of brand. The work presents an architecture for this data market as a facilitator of data services across the motor vehicle sector. The data model is open and harmonized, which allows brand-independent aggregation of generic data and datasets. They applied AutoMat concepts and the prototype to weather forecasting and road quality use cases in non-automotive applications.

Secure Data Exchange (SDE-Secure Data Exchange) (Gilad-Bahrach et al., 2017) aims to ensure the exchange of data between different platforms in the cloud market. The system must meet the requirements of: leveraging existing cloud storage infrastructure; align with existing incentives for cloud services; and use trust models that reflect the current reality of cloud services. The cloud market, from trusted providers, needs a general encryption solution for the owners of various data stored in different clouds, allowing encrypted private data to be available in a cloud without collusion and in an honest way, through an assessor who intends to engage in a security assessment of data belonging to a subset of data owners across platforms.

Jang, Park, Lee, & Hahn (2018) propose three hierarchical levels of the big data market from multiple IoT data sources, through a competitive big data market model, multiple data sources, a service provider and customers. It starts with the service provider that gathers the data from multiple sources and provides key insights from accurate data as a service to the customers, which determines the optimal data acquisition from multiple data sources, with the defined budget constraint. The multiple data sources independently track the action of the service provider through bidding prices. Analytical results demonstrate that the proposed approach guarantees a single break-even point that maximizes returns for all market participants,

The Wibson Decentralized Marketplace (Travizano et al., 2018) aims to empower individuals to securely monetize their personal data. This blockchain-based marketplace provides individuals with a way to securely and anonymously sell information in a trusted environment. The combination of smart contracts and blockchain allows data sellers and buyers to transact directly, while maintaining the ability for individuals to remain anonymous.

4. Mobility Ecosystem

The mobility ecosystem, in a smart city, involves managing demand through a selective capacity of supply, adopting a strategy of system efficiency, in the sense of reducing the need for travel, with the efficiency of travel, seeking to improve or maintain the environmentally sustainable environment, with transport efficiency and improved energy efficiency.

The core elements of mobility include inclusive smart cities, new business, innovation, technology and data models, with infrastructure and connectivity, new vehicles and more autonomous vehicles, through efficient supply chains, supported by energy efficiency, charging optimization, and in increasing forms of active mobility.

The six main goals of "smart" mobility are grouped into the following categories: reduce pollution, reduce traffic congestion, increase people's safety, reduce noise pollution, improve transfer speed and reduce transfer costs. The four main groups of initiatives are classified by: public transport companies and organizations; private companies and citizens; public bodies and local governments; and the combination of actors carrying out integrated initiatives. (Benevolo et al., 2016)

Public policies prove to be adequate if "fueled" by data and information on the functioning of the city, knowledge of the relationships and interdependencies of the different ecosystems and actors, and the participation and involvement of citizens.

The creation of infrastructure should be geared towards the population, be supported by the shared economy, low-carbon energies, new technologies, with more efficient intermodal passages and secure, reliable, scalable, flexible, efficient and integrable ticketing and payment systems the city's products and services.

In this context, transport must adapt to changes to support the mobility needs of people and goods. Information systems seek more efficient operation through automation, security and redundancy.

Security tries to minimize the weaknesses of systems against physical (security), logical (cybersecurity), and accidents and unintentional actions (safety). The methods essentially involve hazard identification, risk assessment and damage control.

City infrastructures and services are more interconnected, enabling monitoring, control and automation. Public and private transport systems integrate and access location, time and traffic data, improving their efficiency in intermodality, public safety and disaster recovery. The balance between the benefits of an efficient city and a city that is more watched and listened to and the risks has a central element, the guarantee that rights and freedoms are protected. (Elmaghraby & Losavio, 2014)

The usability aspects of the applications allow users to find information about the services offered. The metrics and methods used to evaluate the usability of applications related to public transport systems indicate that satisfaction, effectiveness and efficiency are the most used usability metrics. Survey appears to be the most used usability method among investigators, followed by field testing and interviewing in the development of usable public transport systems. (Hussain et al., 2017)

4.1. Systems in passenger rail transport

In the mobility ecosystem, in smart cities, public rail passenger transport has a binding and structuring role in the design of transport flows and in the great capacity to move people in the urban space. Urban rail transit in cities, especially in regions, metropolises and megacities, plays a central role in daily travel and with ever-increasing passenger flows.

The future of rail opportunities for energy and the environment [51]indicates that nearly 200 cities worldwide have metro systems, the combined length exceeds 32,000 kilometers, with light rail systems adding 21,000 kilometers in length in a further 220 cities. This study presents several trends in rail transport that go through:

- Rail transport is one of the pillars of passenger mobility and freight transport.
- The railroad is one of the most efficient and low-emission modes of transport, heavily dependent on electricity.
- The majority of rail networks are located in India, China, Japan, Europe, North America and the Russian Federation, and light rail and metro networks operate in major cities around the world.
- The future of rail transport depends on responding to growing transport demand and increasing pressure from competing transport modes.
- Annual investment in rail infrastructure increases to $315 billion by 2050, based on current projects and various stages of construction and planning, and forecasts strong growth in high-speed rail networks.
- Global rail electricity usage will reach nearly 700 TWh by 2050.
- The two categories, urban rail and high-speed rail, predict great global benefits.

Drivers for urban transport (LRV- Light Rail Vehicles) [52] until 2021 are presented in the following table:

Table II.18- Light rail acquisition trends

drivers for urban transport	Trend until 2021
Urbanization and demographic changes	Strongly increasing
Selection of public transport systems	Increasing
Infrastructure development and upgrade new	Increasing
Replacement acquisitions	Constant
Investment funds	Strongly increasing

The UNIFE report[53] indicates that the rail sector will continue its growth driven by demand, innovation and legislation, and that:

- This industry will continue to grow.
- All continents and all product segments will continue to grow.
- Megatrends continue to drive demand for rail transport.
- Digitization is a new opportunity.

---

[51] The Future of Rail - https://www.aktuellhallbarhet.se/wp-content/uploads/2019/01/the-future-of-rail.pdf, accessed on 30-04-2019.
[52] LRV - https://www.sci.de/fileadmin/user_upload/MC_Studien_Flyer/Flyer_MC_LRV.pdf, accessed on 30-04-2019
[53] UNIFE - https://www.rolandberger.com/publications/publication_pdf/roland_berger_world_rail_market_presentation_final.pdf, accessed on 30-04-2019.

The 2019 UIC (International Union of Railways) report [54] presents the future of the rail transport system in Europe, which will be focused on the customer, in a shared vision, with the integrated mobility system, at the best Value in terms of quality and costs, end customers receiving a quality service, transport to respond flexibly to changes in demand and operating conditions, being a safe, comfortable and sustainable way in terms of energy. This industry will continue to attract talent and innovation.

The RSSB report [55] recommends the application of the BREEAM (Building Research Establishment Environmental Assessment Method) for sustainable assessment for all new station and car park developments and for major renovations to minimize life-cycle carbon. Infrastructure has long lifecycles and is tightly controlled for safety, performance and customer satisfaction, hence opportunities for early carbon reductions in new projects should be leveraged. This tool is useful for assessing the environmental performance of development, although it does not address the broad issues of sustainable development, geographic location suitability of solutions, and long-term environmental footprint. (Sewell & Fraser, 2019)

In these systems, stations and transfers with other modes are critical points for the disturbance of the rail network. Lu & Lin (2019) propose the application of a unique location-based accessibility approach to the vulnerability analysis of the urban rail transit network. The accessibility method that analyzes the vulnerability of the road network is measured by station interruptions, connection and line failures, based on the combination of the accessibility method and the graph theory approach. As a result, the characteristics of the flow of rail passengers, changes in travel costs, alternative modes of transit, the measurement of consequences on the rail network and the implications for decision-making in the disturbances in the rail network are explained.

Poor maintenance engineering performance is a major cause of poor service delivery and business performance. (Fourie & Chimusoro, 2018). Maintenance engineering performance is measured using various parameters, which include the organization's ability to upgrade and replace the fleet, spare parts availability, skills availability, information flow, and information technology.

---

[54] UIC - https://uic.org/europe/IMG/pdf/2019_uic_railway_technical_strategy_europe.pdf, accessed on 2019-04-20
[55] RSSB - https://www.rssb.co.uk/Library/improving-industry-performance/Rail-Industry-Decarbonisation-Task-Force-Initial-Report-to-the-Rail-Minister-January%202019.pdf, accessed on 30-04-2019.

Maintenance is a critical aspect in providing an urban passenger transport system. The maintenance process must be integrated and combined with the "smart" maintenance decision method, based on a decision tree induction strategy to identify the equipment class and rules that led to similar failures. This "key" is defined as the priority for repair, prevention, by predicting possible breakdowns, in order to establish the maintenance program and decision-making. (M. Zhang, 2017)

Liu & Sun (2018) analyze the applications of BIM (Building Information Model) in the construction of the railway system. Local space is limited, resource allocation is complex and often on a short timescale, and complex engineering issues in underground interventions are numerous. BIM technology allows three-dimensional visualization, parameterization, virtual simulation, information (costs, technical options, dimensioning, compatibility, etc.) that accompany the entire design, review, construction and maintenance process. The construction of the metro platform must be based on BIM and Cloud technology, with the use of cameras and sensors to achieve electronic integration, dynamic control of the operation and maintenance of facilities and underground spaces, to improve operation and maintenance. .

Safety is another critical aspect to consider when designing the construction, maintenance and operation of an urban rail transport system. The security risk management processes in the URT (Urban Rail Transit), from the design stage, present indirect security risk factors, such as management defects, direct and participant security risk factors. The proposed model, through the oriented text search method (J. Li et al., 2018), states that for each accident a standardized description of the corresponding accident information is carried out, which allows the accumulation of data and risk analysis. The risk analysis and assessment methods contain the safety risk information integrated into the accident descriptive model using accident points, to be managed and controlled.

In general, URT systems represent challenges to understand cyber threats and their impact, in order to prioritize investments and robustness efforts.

The example of the SECUR-ED project [56] provides guidance for safety organizations in charge of public transport, in particular operators, on daily operations, optimized procedures and vision of future safety systems.

Song, Li, List, Deng, & Lu (2017) present a method to study the vulnerability factors of a URT (Urban Rail Transit) system, based on the analytical hierarchy process (AHP-Analytical Hierarchy Process) and on interpretive structural modeling (ISM-Interpretative Structural Modeling) whose results provide information for the decision-making, with proactive strategies and policies to reinforce security and promote the sustainable development of urban space. In the following figure, adapted from table 1 of Song, Li, List, Deng, & Lu (2017) the 6 dimensions that group the 21 vulnerability factors are presented.

Table II.19 - Vulnerability Factors in URT

Dimension	Factors
Individual (B1)	Individual technical capacity (C1)
	Individual safety awareness (C2)
	Individual subject (C3)
	Individual workload and stress (C4)
	Individual physical and physiological state (C5)
Equipment / facilities (B2)	Equipment / installation condition (C6)
	Equipment / facilities performance (C7)
	Protection of equipment / facilities (C8)
Environment (B3)	Natural environment (C9)
	Social environment (C10)
	Operating environment (C11)
Management (B4)	Investment in safety (C12)
	Education and training (C13)
	Rules and Regulations (C14)
	Organizational Structure (C15)
Structure (B5)	Station layout (C16)
	Network topology (C17)
	Equipment / installation interdependence (C18)
Emergency (B6)	Emergency management plan (C19)
	Execution of emergency response (C20)
	Emergency support system configuration (C21)

Adapted from table 1 of Song, Li, List, Deng, & Lu (2017)

Human resources are the cornerstone of these systems. Assessing workloads leads to more collaborative and sustainable practices that can minimize the risk of overload and failure. The systems approach to predicting and measuring the workload of rail traffic

---

[56] SECUR-ED Project - http://www.secur-ed.eu/wp-content/uploads/2014/12/SECUR-ED_White_Paper_Final.pdf , accessed on 30-04-2019.

management systems, through the In2Rail project (Evans, 2017), explores diverse issues around core workload principles, through increased collaborative work styles and shared responsibilities, with greater flexibility, more configurability to support the information requirements for each specific role and support each role during different scenarios. The In2Rail toolset was intended to make traffic management system workload forecasts with a higher level of confidence than existing workload measurement techniques to support the continuous development of people, processes and technology.

4.2. Information technologies in public transport systems

Information technologies have supported public transport operators in improving safety and the quality of the service provided to their customers. Infrastructure security is geared towards promoting risk management methodologies with the assessment of system vulnerabilities and the design of procedures, best practices and technologies to protect against cyber-attacks. The assessment becomes permanent and places the systems under continuous monitoring, reinforced by the most complex and interdependent technological infrastructure.

The information system of an urban passenger transport system can be subdivided into four supporting information systems, as follows:

- Control and Management;
- Exploration;
- Rail Traffic Management;
- Rolling Stock.

In the following points, the set of systems that can constitute each of the support information systems will be detailed.

4.2.1. Control and Management Support Information System

This system covers the set of BackOffice, management, control and planning applications that provide transport companies with the basic functions to meet legal, tax, reporting, analysis and evaluation obligations.

In this domain, there are applications for management (ERP-Enterprise Resource Planning), human resources management, operational planning (programmed theoretical offer), allocation of vehicles and stopovers, event/incident management, maintenance management, document management and quality, the basic infrastructure, physical and logical security, service quality indicators from the datawarehouse and BI (Business Intelligence) tools.

These applications and data flows are described in the following figure.

Figure II.8 - Information System to Support Planning and Management

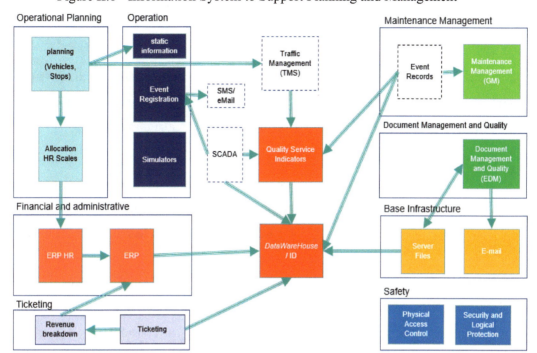

Operational planning, referring to vehicle services and human resources scales, presents the traffic graphics, theoretical services and the resolution of incompatibilities, and the allocation of human resources to theoretical scales. The static information is made available by the timetable generation/formatting application.

The ERP application can include accounting, purchasing, commercial, financial and logistics. From the ticketing system, the value of revenue sharing in an intermodal system is received. The application of human resources allows the processing of salaries, namely with the allocation of scales.

In operation, through the event log, incidents and corrective maintenance actions are managed. Driving simulators by vehicle type and maintenance simulators are also available.

The maintenance management application allows you to plan preventive maintenance, manage maintenance resources, manage corrective maintenance works, manage stocks and manage rotables.

Document and quality management makes it possible to control the life cycle of documents and the quality certification processes.

The base infrastructure presents the numerous servers for managing the domain, network, file sharing, email, telephone, networking, DNS and many others necessary for the basic functions, namely physical and logical security.

The quality of service indicators are the result of the set of options taken in the different domains. Its calculation is based on data from the various measurement systems and procedures available in the datawarehouse and using BI calculation tools.

Tyrinopoulos & Aifadopoulou (2008) present a methodology for quality control of services provided to public transport passengers, which include characteristics such as safety, performance, punctuality, accessibility and efficiency. The proposed methodology has seven main categories: safety, comfort and cleanliness; information and communication with passengers; accessibility; terminals and stopping point performance; lines/routes performance; general elements of the public transport system, namely the tariff and ticketing system, timetables; and composite indicators as a result of indicators from the five previous categories, for a consolidated picture of performance or satisfaction/dissatisfaction.

The perceptions of service quality and loyalty influence the user satisfaction of public transport users. (van Lierop & El-Geneidy, 2016) The Table II.20, based on table 6 of van Lierop & El-Geneidy (2016), presents a summary of the effectiveness of the implementation of strategies and the impact on the three groups that use the system: captive users, users by choice and captive users by choice.

Table II.20 - Summary of the effectiveness of strategies

Strategies	Captive	Choice	captive by choice
Bus service quality	strong impact	medium impact	medium impact
Metro service quality	medium impact	strong impact	medium impact
Reliability	strong impact	medium impact	strong impact
Safety	strong impact	medium impact	medium impact
Training	medium impact	strong impact	strong impact
Cleaning	medium impact	medium impact	strong impact

Adapted from table 6, from van Lierop & El-Geneidy (2016)

These strategies reveal the need to intervene in all of them given their strong impact on at least one of the user groups.

4.2.2. Exploration Support Information System

These systems are concentrated at stations, car parks and intermediate routes. The stations have systems in order to support passengers in accessing transport in an easy, fast and comfortable way.

These systems support a wide range of functions that include: the station management system that manages the elevators, escalators, the fire detection and suppression system, pumping, smoke removal, opening and closing doors, lighting, energy and communications; the public information system (SIP) that guides passengers to the respective docks with information on services, destinations and waiting times and, in the event of a disturbance or emergency, guide their movement at the stations; the multimedia system allows the broadcasting of corporate channels with the possibility of inserting the SIP; the video surveillance system makes it possible to record and monitor images to support the exploitation and security of people and goods, using image processing and analytics to detect various security events; the energy management system makes it possible to manage the traction electrical network, supporting the other systems, switching, redundancy and rescued energy.

These systems and the traffic management systems are normally centralized in an operations control center for the different lines.

The Figure II.9 presents the Exploration Support Information System with the various relationships and data flows that are established between the various applications.

Figure II.9 - Exploration Support Information System

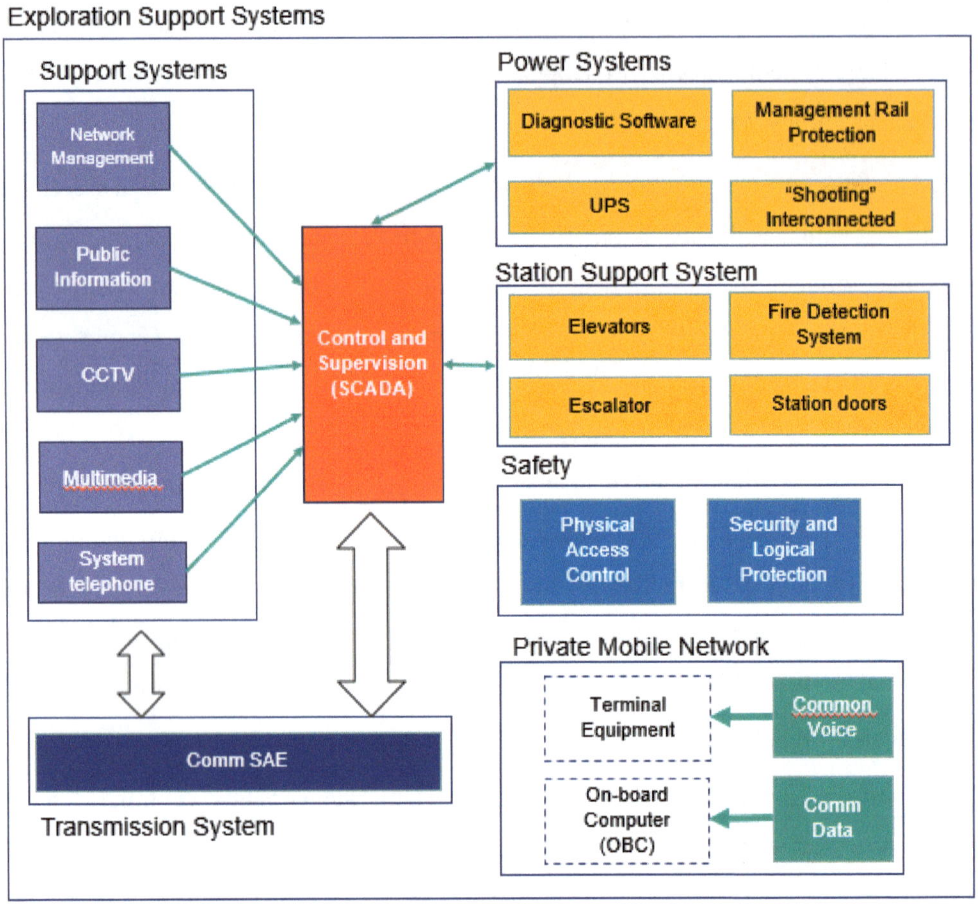

The transmission system integrates the set of telecommunications systems and support modules necessary for the operation of the services provided over telecommunications networks, namely, the application of network management and support for the operating support systems.

The SCADA (Supervisory Control and Data Acquisition) system, for the supervisory control and data acquisition, of a public infrastructure presents a security risk, with complex, interconnected and often old systems. (Temple et al., 2017). This system has two main functions, the control and monitoring function and the supervision and actuation function. These functions include the management system for the overhead contact line, the energy network (low voltage, medium voltage and, if applicable, high voltage), lighting, escalators, elevators, ventilation, smoke removal, pumping, fire detection and suppression, rescue power systems, industrial battery chargers, temperature control, ventilation system speed, measurements (voltage and amperage), URTs, level crossings (LC), alarms and events and the various auxiliary services.

Energy management includes medium voltage protection systems, diagnostic software and relay configuration, control by URTs (Remote Terminal Unit), protection, command and control systems supported by the SCADA system, rail and trip system interconnected through programmable automata for network switching and redundancy.

4.2.3. Information System to Support Rail Traffic Management

The safety and accuracy of these systems are related to the safety and efficiency of rail transit operation and the safety of passengers' lives. The metro signaling system safety certificate must be obtained in order to carry out the passenger transport operation, respecting tests, standards, procedures, training and exhaustive verifications of its compliance.

Dynamic models and algorithms for rail traffic management allow optimizing rail timetables, making operations more robust and resistant to deviations. The daily plan can be adjusted trying to keep operations viable and avoid the spread of delay. Online rail traffic rescheduling problems have dynamic and non-deterministic aspects. The initial static scheduling contains the theoretical offer with the theoretical probability of the scheduling problems, the dynamic rescheduling of the rail traffic allows to minimize the uncertainty for the future states (Corman & Meng, 2015).

Computing agents allow the development of large-scale distributed systems in a dynamic environment. Traffic management systems manage geographically distributed transport systems. The techniques and methods of agent and multi-agent systems are applied in this domain in modeling and simulation, dynamic routing, congestion management, dynamic traffic control, seeking to resolve critical issues such as interoperability, flexibility and extensibility. (Bo Chen & Cheng, 2010)

Mazzarello & Ottaviani (2007) present an architecture for the implementation of an advanced system of Traffic Management (TMS-Traffic Management System), in real time, capable of optimizing the confluence of traffic in railway networks with different signaling systems, seeking to solve the increase in traffic intensity and the complexity of the rail system.

Also Li, Yang, & Gao (2015) seek solutions for the coordinated control of train movements based on a multi-agent model. The movement of an ordered set of vehicles operating on a railway line is modeled by a multi-agent system, in which each vehicle communicates with adjacent vehicles to adjust the speed and can follow the desired speed, the distances between vehicles are stabilized within a safety area.

The railway is driven by mobile communications services, which evolves into an infrastructure where vehicles, passengers, travelers and goods are more interconnected and can provide finer mobility, comfort and greater safety. Wireless, high-speed data connectivity evolves into high-data, high-definition services enabling new passenger, security and surveillance services (Ai et al., 2015).

The predictive "intelligence" for a rail traffic management system (Roberts et al., 2017) must foresee the increase in demand for transport systems, translated into greater efficiency and more "intelligent" mobility, in autonomous means of transport that will allow the evolution of the technology of the "intelligent" transport system (ITS- Intelligent Transport Systems) and of the cooperative ITS (C-ITS- Cooperative Intelligent Transport Systems[57],). Cooperative ITS enables more benefits in ITS services and applications, which communicates and shares information from ITS points, to improve efficiency, comfort, sustainability and safety

Tge Table II.21 based on figure 1 of the UITP summary[58] on automation trends, introduces the four degrees of automation. There are several degrees of automation (GoA-Grades of Automation). These are defined according to the basic functions of vehicle operation which are the responsibility of the personnel and the responsibility of the system itself. For example, an Automation Degree of 0 would correspond to on-site and spot operation. Automation grade 4 refers to a system in which vehicles run fully automatically without any operational staff on board.

---

[57] CITS - https://www.itsstandards.eu/cits, accessed on 7/10/2019
[58] UITP - Metro Automation - https://www.uitp.org/sites/default/files/Metro%20automation%20-%20facts%20and%20figures.pdf, accessed on 7/9/2019.

Table II.21 - Degrees of Automation

Automation Degree	Vehicle operation type	Setting up the vehicle in motion	vehicle stop	door lock	Operation in case of interruption
GoA 1	ATP with Conductor	Conductor	Conductor	Conductor	Conductor
GoA 2	ATP and ATO with Conductor	Automatic	Automatic	Conductor	Conductor
GoA 3	without driver	Automatic	Automatic	vehicle support	vehicle support
GoA 4	UTO	Automatic	Automatic	Automatic	Automatic

Automatic Vehicle Protection (ATP-Automatic Train Protection) is the system responsible for basic safety, by preventing collisions between vehicles, crossing red lights and exceeding speed limits.

The automatic vehicle operation (ATO-Automatic Train Operation) allows automatic or partial vehicle driving and driverless functions, when performing driving functions, although it does not close the doors.

Automatic Vehicle Control (ATC-Automatic Train Control) automatically performs normal signaling operations, such as route configuration and vehicle regulation. The ATO and ATC systems work together to keep a vehicle within the tolerance defined in your schedule. The combined system adjusts operating parameters marginally.

The implementation of Unattended Train Operation (UTO) systems allows operators to optimize vehicle operation time, increasing the average speed of the system.

This growing automation is reflected in greater flexibility in the operation, in greater operational security, with an increase in the quality of service and an improvement in the conditions of financial viability.

The Next Generation Vehicle Control (NGTC-Next Generation Train Control) program studies the similarities and differences between the European Train Control System (ETCS)[59] and the communication-based vehicle control systems CBTC (Communications-Based Train Control[60]) (Gurnik, 2016). ETCS is a vehicle control system developed for the main European railway lines, while CBTC systems were developed separately for urban rail vehicle control systems. ETCS seeks to promote

---

[59] ERTMS/ETCS - http://www.railwaysignalling.eu/wp-content/uploads/2016/09/ERTMS_ETCS_signalling_system_revF.pdf , accessed on 13-07-2019.
[60] IEEE. (2004). IEEE Standard for Communications-Based Train Control (CBTC) Performance and Functional Requirements. https://ieeexplore.ieee.org/document/1405808

interoperability across lines. The ERTMS / ETCS (or just ETCS) L2 signaling system provides the information for safe driving, regarding the effects of actions, changes in line conditions and the activation of emergency braking if the speed exceeds the maximum allowed.

Communications-based vehicle control systems (CBTC) are subway signaling platforms, which coordinate and protect vehicle movements within a station's rails and between different stations. On these CBTC platforms, the main function is performed by the system (ATS-Automatic Train Supervision), which automatically chooses and directs vehicles on the network, and avoids deadlock situations.(Mazzanti & Ferrari, 2018)

The CBTC system is a complex distributed system in an open and dynamic environment. The main factor that constrains the performance of the system and the degree of improvement is its complexity. Mainly analyzed system complexity from two aspects of structure and function implementation(T. Chen et al., 2018). The following figure is adapted from table 2 of Chen et al. (2018) which describes the main functions of the system.

Table II.22 - Main functions of CBTC

Kind of equipment	equipment name	function name
On board	On-board ATP	Vehicle location / vehicle speed determination
		About speed protection
	ACT on board	automatic vehicle driving
side of the way	Zone Controller (ZC-Zone Controller)	vehicle management
		Calculation of Movement Authorities (MA)
	Computer Interlocking (CI-Computer Inter-locking)	route interlock
	Automatic Vehicle Supervision System (ATS-Automatic Train Supervision System)	Automatic vehicle supervision and regulation

The urban rail transport system can be described as a metric space in which topological spaces are introduced to form movement authority and trajectory (track occupancy). Security rules are verified by performing a series of verification calculations based on the topology. (Haifeng Wang et al., 2018)

The Figure II.10 presents a scheme of a traditional traffic system, involving ATP and ATR as protection and regulation systems.

Figure II.10 - Generic Traffic Management Support Information System

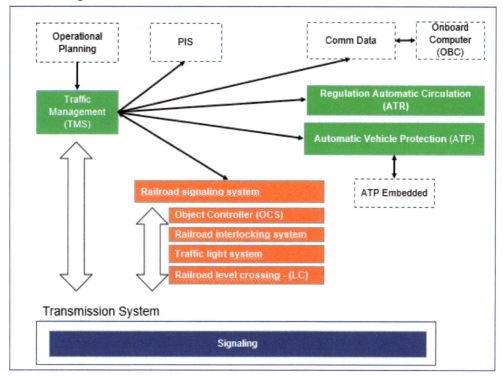

The signaling system (SIG - Signaling System) is generally composed of: the vital interlocking system based on a computer and on the interlocking processing unit (IPU - Interlocking Processing Unit); the Object Controller System (OCS); traffic lights, in the case of systems that interact with car traffic and pedestrians; the traffic management system (TMS - Traffic Management System); automatic train regulation (ATR - Automatic Train Regulation); and the on-board computer (OBC - On-Board Computer). In addition to other protection systems adopted (ATP, ATO or UTO).

The signaling systems use the beacons on the road, being a passive device that is energized when a vehicle passes and that communicates with the beacon transmission module (BTM - Balise Transmission Module) connected to the vehicle through telegram messages. This equipment is critical and its interface can be exploited to change or manipulate information about telegrams. Security and integrity mechanisms must be reinforced with security protection for telegram messages with AES-CCM (AES-Advanced Encryption Standard, CCM - Cipher Block Chaining-Message Authentication Code) approaches.(Dworkin, 2007) and HMAC (Hash Message Authentication Code) that add protection and data integrity verification to telegram messages received by BTM from beacons. (Guo et al., 2018)

The safety management and signaling system integration model, suitable for the control of basic communication with the vehicle (CBTC-Communication-Based Train Control), in accordance with CENELEC standards[61], aims to ensure security and assessment based on security verification and validation process. The method was applied at all stages of the CBTC system development life cycle, to monitor and control each activity in the life cycle and to evaluate each document in the system development process and ensure the traceability of the relevant documents and test all system functions (Yan et al., 2017).

4.2.4. Rolling Stock Support Information System

The rolling stock support systems make it possible to manage, control and operate the various subsystems installed in vehicles and which are related to their movement. Functions include vehicle management and control, data and voice communications, traction energy management, lighting, HVAC, doors, passenger information system (PIS), video surveillance system, passenger counting , multimedia systems, protection and security systems, among other systems. The following figure describes the main vehicle control and command subsystems.

---

[61] IEC 62278-2002 Railway applications - specification and demonstration of reliability, availability, serviceability and safety (RAMS)
IEC 62279-2002 Railway applications - communication, signaling and processing systems - software for railway control and protection systems
IEC62425-2007 Railway applications - communication, signaling and processing systems - safety-related electronic systems for signaling

Figure II.11 - Rolling Stock Support Information System (MC)

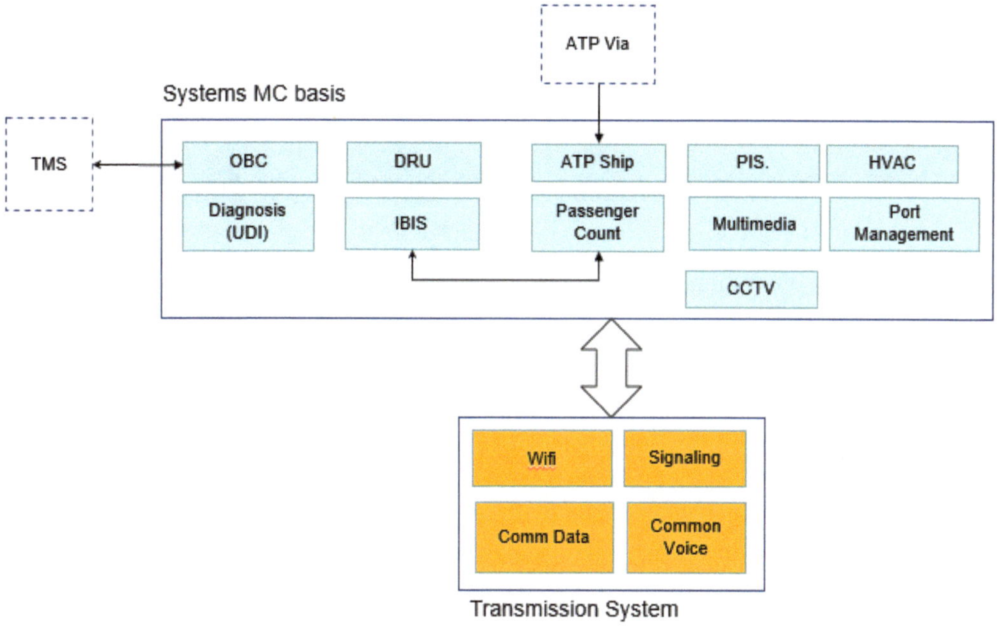

The rolling stock support information systems are generally composed of: the on-board computer (OBC - On Board Computer); the embedded information system (IBIS - Integrated On-board Information System); the event recording unit (DRU - Data Recording Unit); the HVAC air conditioning system; the multimedia system; the voice communications system; the data communication system that allows the connection to the TMS; the passenger information system (PIS - Passenger Information System); the central computer (VTCU - Vehicle Train Control Unit); the traction control unit; vehicle's main bus; and the diagnostic monitor.

4.3. Ticketing System

This system is dealt with in greater detail since it will be on it that the proof-of-concept works, within the scope of the thesis, will be carried out and that it is intended to be able to validate the proposed model in the alarmistic and data flow control components.

The ticketing system generally comprises a sales network (local and decentralized), a validation network and a central database that centralizes all sales, validation and monitoring data, and replicates them for each of the operators. In more detail we will have the following modules or function blocks:

- Ticketing equipment which may include vending and charging equipment, validation equipment, inspection or vending equipment, and local controller equipment and operator controllers. There are also decentralized ways of selling, validating and e-ticketing.
- The BackOffice application allows the management of all ticketing functions and each of the operators of their single-modal ticketing data. The main functions include the commercial management of sales and loading, revenue control, incorporation into accounting, calculation of the origin/destination matrix, registration of validations, inspection data and data from sensors and alarms.

The following figure summarizes the general architecture of the ticketing system.

Figure II.12 - General Architecture of the Ticketing System

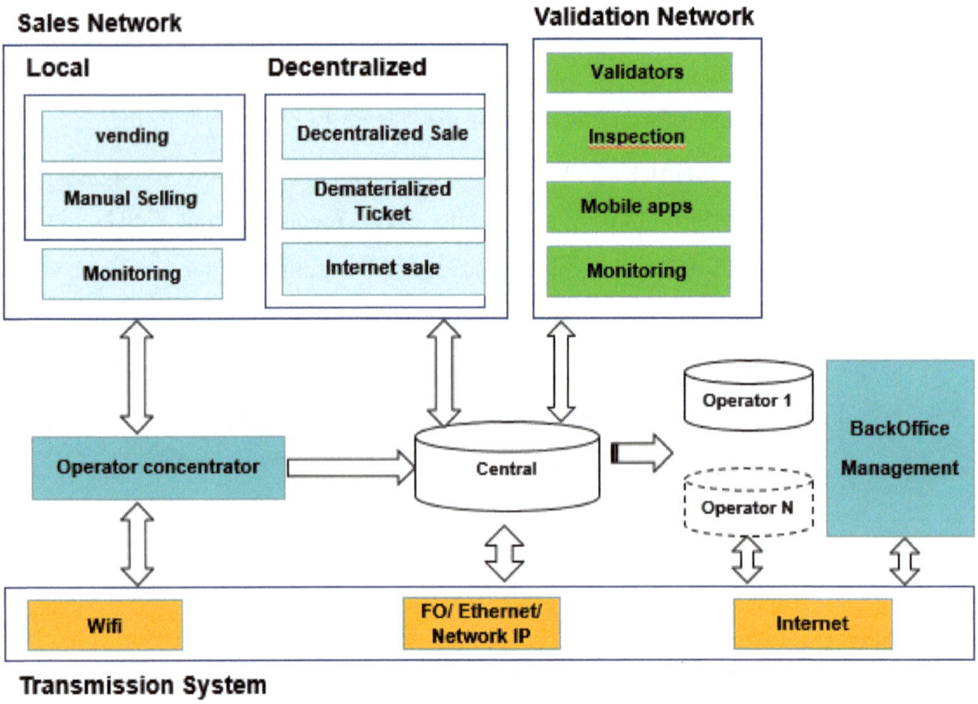

The base architecture shown in the previous figure allows several data flows for transactional recording, although consolidated in a central database. The new dematerialization solutions are also present and work in a decentralized way,

Ferreira, Nóvoa, Dias, & Cunha (2014) propose a e-ticketing solution in public transport using mobile devices to minimize the investment cost of public transport operators and maximize customer acceptance.

In the same sense, Mallat, Rossi, Tuunainen, & Öörni (2007) conducted a survey for the adoption of the mobile e-ticketing service in public transport, which suggests the compatibility of the mobile ticketing service with customer behavior, mobility and contextual factors, including budget constraints, the availability of other alternatives, and time pressure on the service usage situation, which are what determine its adoption and which can be integrated into traditional adoption models.

Customer acceptance and examples of mobility technology adoption (Bongaerts et al., 2017) part of new challenges with: rising energy costs; the internet of things (IoT); blockchain technology; the change in communication channels; and the biggest competition. The challenges in the mobility business move to the level: customer / consumer / user due to demographic trends, urbanization, car sharing, living in a digital world, simplicity / convenience, real-time and ubiquitous solutions; from the supplier due to rising energy costs, the internet of things/big data, the fit for mobile devices, transparency and new competition outside the industry; and governance due to the rising cost of infrastructure, global warming, healthcare and security. The development of blockchain technology could result as a new impetus for more connected mobility,

Ticket lifecycle management should make it possible to combine multiple providers of goods and services. The difficulty lies in an architecture or a unified technological solution for the management of e-commerce as a result of the transaction with complementary services combined with the management of the electronic ticket that relate to different systems, applications and interfaces of the transport sector and other services that can relate (Bumanis et al., 2017).

Nair, Pawar, Tidke, Pagar, & Wani (2018) propose solutions based on mobile devices with transport location data, online e-ticketing, cash payment ticket generation and QR code validation (Quick Response).

Sheikh, Khapekar, Kumar, & Kumar (2018) present a review of the techniques of the e-ticketing system for the implementation of the electronic ticket. This review concludes that although RFID, GPS, WIFI technologies and the AZTEC code are used, it is the two-dimensional QR code (comparison of codes[62]) quick response that has distinguished itself for being easy to use and versatile for efficient e-ticket system.

---

[62] 2D Barcode - https://www.tec-it.com/en/support/knowbase/barcode-overview/2d-barcodes/Default.aspx , accessed on 07-13-2019

Kazi, Bagasrawala, Shaikh, & Sayyed (2018) present an "intelligent" e-ticketing system for public transport that aims to respond to needs such as undue waiting time at stops, reimbursement for non-compliance and the lack of seats for passengers. The objective is an agile and smooth ticketing that allows the automatic allocation of the passenger's seat with the reservation of the digital ticket, with payment without using cash, in which the user checks the availability of seats, books tickets, obtains the seat automatically through an efficient algorithm and according to the expected waiting time.

Direct fundamental aspects are present in ticketing systems, such as the availability of the ticketing service and pricing, and indirect aspects, such as the quality of the transport service provision, punctuality and regularity, compliance with the offer, maintenance of equipment and infrastructure, cleaning, information to the public and communication with the customer.

In this work, it will be mainly focused on the availability of the service, a result of the system's alarmism and, consequently, the fulfillment of timely maintenance.

III. EMPIRICAL STUDY

III. EMPIRICAL STUDY

1. Methodology

At this point, we seek to obtain information to choose the most appropriate scientific methodology to develop the proposed work.

The following table summarizes chronologically the contributions considered most relevant to the development of the DSR (Design Science Research) methodology in the IS/ICT domain.

Table III.1 - Contributions to the DSR methodology

Year	Contribution	References
1992	It defines an information systems design theory (ISDT- Information Systems Design Theory) to produce more effective information systems.	(Walls, Widmeyer, & El Sawy, 1992)
2002	It defines the DS (Design Science) in ICTs with the intervention in the representational world defined by the hierarchy of concerns following the semiotics. The complementary nature of the representational (internal) and real (external) environments provides the basis for articulating ontological and epistemological foundations.	(Purao, 2002)
2004	The predominant research tends to be description-oriented, based on the "explanatory science" paradigm, resulting in Organization Theory. Relevance can be mitigated with prescription-based research, on the "design science" paradigm and on Management Theory. , where the research outputs would be "field-tested and grounded technological rules." The nature of these rules is discussed, as are the research strategies that produce them.	(Aken, 2004)
2006	The DS is defined and compared with other research paradigms and presents seven guidelines for understanding, executing and evaluating design research.	(Manson, 2006)
2006	It develops a framework of activities for the interaction of Design Science with research in other scientific paradigms.	(Venable, 2006)
2007	It defines three related cycles of activities: the cycle of relevance, the cycle of rigor and the cycle of central design, recovering the pragmatic nature of DS.	(Hevner, 2007).
2007	DS research must meet three objectives: to be consistent with previous literature, to provide a nominal process model for conducting DS research, and to provide a mental model for the presentation and evaluation of DS research in the SI. The DS process has six steps: problem identification and motivation, definition of goals for a solution, design and development, demonstration, evaluation, and communication.	(Peffers et al., 2007)
2009	Soft Design Science (SDS) looks for new ways to improve human organizations, in the social aspects, with activities of design, development, instantiation, evaluation and evolution of a technological artifact. The SDS approach combines the common DS research process (design, construction-artifact, evaluation) with the iterative methodology of Soft Systems, in a design-oriented action research method.	(Baskerville et al., 2009)
2011	They present a DS roadmap for planning, executing and communicating DS research, with other constructive criticisms, improvements and extensions, and with a broad coverage of DS	(Alturki et al., 2011)

Year	Contribution	References
	research aspects and activities.	
2012	They discuss the semantics of design theory from an epistemological view of the framework, relating it to an idealized cycle of scientific research and seek to demonstrate the potential of the DSRIS (Design Science Research in Information Systems) framework.	(Kuechler & Vaishnavi, 2012).
2013	DSR (Design Science Research) has two dimensions: the state of existing knowledge in the problem domains and the solution to the research opportunity under study. The DSR reporting scheme has similarities to conventional publishing standards in that it replaces the DSR artifact description in the traditional results section.	(Gregor & Hevner, 2013)
2015	It describes a methodology for carrying out the DSR, through a comprehensive guideline for conducting research and deepens the theory of DS (Design Science) and different types of theory.	(Vaishnavi et al., 2015)
2017	They compare six DSR methodologies through an existing information systems development methodology comparison framework, to support DS researchers in choosing the appropriate and best-adapted DSR methodology. The six methodologies are: SDRM (Systems Development Research Methodology); DSRPM (DSR Process Model); DSRM (Design Science Research Methodology); ADR (Action Design Research); SDSM (Soft Design Science Methodology); and PADR (Participatory Action Design Research).	(Venable et al., 2017)

Based on the summary of the various contributions considered, the ones that will be analyzed are detailed below, in the sense that they may become the most appropriate methodology to respond to the objectives of this work.

Information Systems Design Theory (ISDT) should be a prescriptive theory integrating normative and descriptive theories with design paths to produce more effective information systems. (Walls et al., 1992). This theory presents the seven characteristics that differentiate design theories: they must deal with objectives as contingencies; it can never involve pure explanations or predictions; are prescriptive; are composite theories that encompass theories of natural science, social science, and mathematics; explanatory theories say "what is", predictive theories say "what will be", and normative theories say "what should be" while design theories say "how or why"; show how explanatory, predictive or normative theories can be used in a practical way; and are theories of procedural rationality.

The vision of the three cycles of DS research presented by Hevner (2007) is related to project activities. The relevance cycle has as inputs the requirements for the research and presents the research artifacts in tests. The rigor cycle presents theories and methods together with domain experience, the knowledge base for the research, and the resulting new knowledge. The design cycle supports the research activity for the construction and evaluation of design artifacts and processes. The following figure, adapted from figure 1 of Hevner (2007), presents the DSR cycles in greater detail and seeks to establish the dynamics that result from possible interactions beyond the cycles themselves.

Figure III.1 - DSR cycles

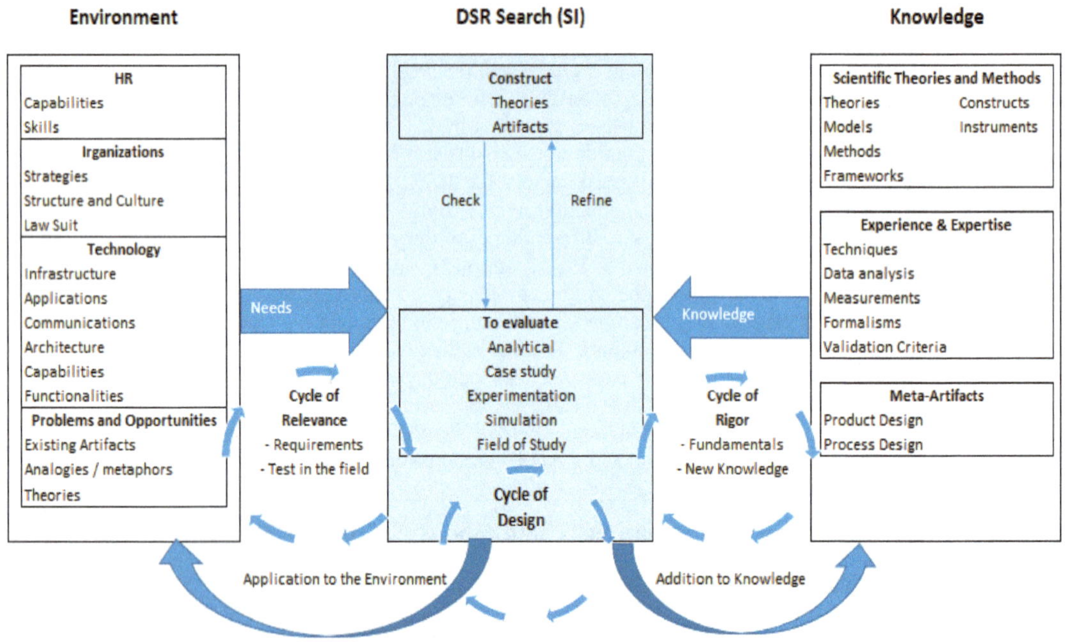

Adapted from figure 1, from Hevner (2007)

Peffers, Tuunanen, Rothenberger, & Chatterjee (2007) demonstrate a methodology for carrying out a scientific research project (DS-Design Science) applied to information systems. The research methodology (DSRM-Design Science Research Methodology) includes principles, practices and procedures for carrying out research. This methodology has three objectives: consistent with previous literature; a nominal process model of the DS search; and a mental model for the presentation and evaluation of DS research. The process is defined in six steps: problem identification and motivation, objective definition, design and development, demonstration, evaluation and communication. The following figure, adapted from figure 1 of Peffers et al. (2007), presents the mentioned steps and the relationships established for the DSRM process and the proposed iterative process.

Figure III.2 - DSRM Process Model

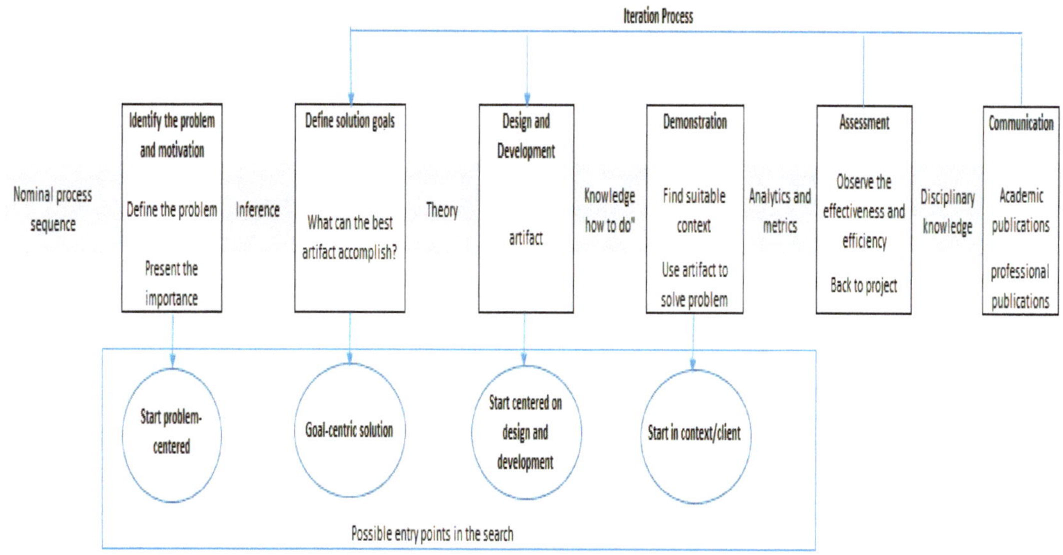

[Diagram: DSRM Process Model showing nominal process sequence with steps: Identify the problem and motivation (Define the problem, Present the importance) → Define solution goals (What can the best artifact accomplish?) via Inference → Design and Development (artifact) via Theory → Demonstration (Find suitable context, Use artifact to solve problem) via "Knowledge how to do" → Assessment (Observe the effectiveness and efficiency, Back to project) via Analytics and metrics → Communication (Academic publications, professional publications) via Disciplinary knowledge. Above: Iteration Process. Below: Possible entry points in the search - Start problem-centered, Goal-centric solution, Start centered on design and development, Start in context/client.]

Adapted from figure 1, from Peffers et al. (2007)

Choosing the most appropriate methodology for DSR (Design Science Research) is an important and critical challenge for researchers. This choice can support or condition a research work and its scientific support.

the work of Venable, Pries-Heje, & Baskerville (2017) presents a framework to compare the six DSR methodologies from the adaptation of the framework (Avison & Fitzgerald, 2006, pp. 597-603) to compare the ISD (Information Systems Development) methodologies, according to the following table, adapted from Table 8 of Venable, Pries-Heje, & Baskerville (2017).

Table III.3 - Framework to compare DSR methodologies

Framework element or sub-element	Description
1. Philosophy	
a. Paradigm	Systems vs. Science, Objectivist vs. Subjectivist Ontology, Positivist vs. Interpretive Epistemology
b. Goals	Possible goals/objectives for DSR methodologies include: increasing relevance; increase the rigor of research; improve (for whom - customer? stakeholders? public?; in what way? effectiveness? effectiveness, efficiency, ethics); emancipation / critical perspective; stakeholder consensus; solve the "right" problem; artifact effectiveness; relationship with the existing literature; practical significance; or theoretical significance.
c. Domain	No specific client, Single client, multiple/group clients, social client
d. Target	Artifact Type: SI/TI, CBIS (Compute-based information systems), ISD method/tool/technique/methodology, product (generally not just SI/IT), process (generally not just SI/IT)
2. Model	What is the basic abstraction and representation mechanism used? (1) verbal, (2) analytical or mathematical, (3) iconic, pictorial or schematic, and (4) simulation
3. Techniques and Tools	What tools and techniques are used in the methodology?

Framework element or sub-element	Description
4. Purpose (DSR activities)	What steps/activities of the DSR process are covered? Activities found in common across DSR methodologies include: (a) Problem Assessment, (b) Design/Framing, (c) Design/Construction, (d) Assessment, and (e) Reflection.
5. Outputs	What are the deliverables at each stage and at the end? (Same as for ISD)
6. Practice	
a. Background	Commercial or Academic?
b. User base	Numbers and types of users of the DSR methodology (use citations as substitutes)
c. Participants	What roles participate and what competencies are needed? Researcher, Customer, User, Other Stakeholders
7. Product	What do methodology buyers get for their money? Software? Training? Documentation? Help service? Consultancy? Etc. (Same as for ISD)

Adapted from table 8, from Venable, Pries-Heje, & Baskerville (2017)

In the context of our work and in detail, we can answer that: paradigms (1a) are science-based, objectivist and positivist; aims at (1b) a new artifact or improvement of existing ones and in the non-customer-specific domain (1c); targets (1d) CBIS, IT or models; within the scope of DSR activities (4), the problem is identified, the objectives of the solution are defined, the design and development of a solution is intended, it is intended to validate and adapt for general use, in the reflection it is intended to carry out communication with the scientific community; outputs an artifact; has access to academia as its practice and researchers and users (evaluators) as participants, and the product will be mainly scientific articles.

These answers to the questions raised in the Table III. and applying the answers according to table 9 (Venable, Pries-Heje, & Baskerville, 2017, p. 7), it appears that the DSRM methodology is the most adequate to the objectives of this work.

In the same sense and using the comparisons between the different research paradigms (Vaishnavi et al., 2015) it appears that this work has design (project-based) as its basic orientation; the ontology of multiple alternatives of contextually situated, socially and technically enabled world states; axiology is based on control, creation, improvement and improvement progress, and understanding, epistemology based on knowledge by doing, in which construction is objectively constrained within a context and iterative circumscription discovers meanings, in approaches or DSR research methods.

Information systems design theory (ISDT) is defined as a prescriptive theory that integrates normative and descriptive theories into design paths aimed at producing more effective information systems. (Walls et al., 1992). The DSRM suggests that one should observe and measure how much an artifact supports a solution to the problem.(Peffers et al., 2007).

A conceptual model (J. Johnson & Henderson, 2002) intended to be the high-level description of the way a system is organized and operated. The model must specify and describe: the main metaphors and analogies used in the design; system concepts aimed at users, which include data objects, domain objects, activities that users create and manipulate, attributes and operations that can be performed; the relationships between these concepts; and mappings of the concepts and domain of activities that the system was designed to support.

Real problems must be properly conceptualized and represented, using appropriate techniques in which the solution must be constructed, implemented and evaluated using appropriate criteria. (March & Smith, 1995)

In the case of the conceptual model for an interactive system there must exist: an idealized view of how the system works; the ontological structure of the system, with objects, relationships and control structures; and the mechanism by which users carry out the activities that the system must support. (J. Johnson & Henderson, 2002)

In summary, scientific design research (DSR), in a problem-centered approach, is considered appropriate for proof of concept that validates the proposed generic data model, supported by blockchain. The DSR methodology followed is based on the six steps foreseen in the DSRM process of Peffers, Tuunanen, Rothenberger, & Chatterjee (2007), using research methodologies based on conceptual models (J. Johnson & Henderson, 2002; March & Smith, 1995) and the design of the prototyping project (Walls et al., 1992).

1.1. Method

The method that supports this scientific work is the DSRM, according to the model of Peffers, Tuunanen, Rothenberger, & Chatterjee (2007), reinforced by the expected approach to be centered on the problem and on the iterative dynamics of the model, which strengthens the way in which the model proposed in the scope of this work can be improved and validated. The proposed generic data model intends to conceptualize the functioning of a smart place, specifically a smart city and that seeks to validate through artifacts that materialize proofs of concept.

The steps follow the 6 steps described in Figure III.2 - DSRM Process Model, with the iterative process proposed in a Problem Centered approach. (Peffers et al., 2007)

1.2. Phases

At this point, the steps referred to in the methodology are briefly described below.

First Stage - Problem identification and motivation

Data management and governance of the growing amount of data generated by a multitude of devices is a technological and management challenge, particularly as the data and the resulting information develop as strategic and decision resources, with characteristics that make them different to govern, the typical ICT artifacts. Controlling data flows, managing the life cycle of data and information has become a critical point in the process of data management and information management and its governance.

Second Stage - Solution Objective

The objective is to develop a generic data model, to support the concept of smart places, particularly smart cities, in order to systematize their actions on data, the control of data flows and data quality, which allow managing data and information, reliably and securely.

The project intends to test the model through the development of an applicational artifact, with a distributed and adaptable design to answer the questions of data flow control, data and information management and its governance, proposing the use of blockchain technology to guarantee the effectiveness of the model.

3rd Stage - Design and development

The artifact aims to test and accomplish the following:

1. Structure a generic data model to support the smart place concept that leads and allows the alignment of the application of data ecosystems with natural ecosystems.

2. Structure the relationships between ecosystems, participants and data that facilitate the use of blockchain technology in data management, security and privacy.

3. Ensure privacy and reliability mechanisms in data management.

The proofs of concept focus on data security and data flow control, on the mobility ecosystem, on the application domain of public rail transport, on the ticketing system and on its data model associated with alarming events that condition the availability of systems and equipment and the correct calculation of the service levels contracted for support and maintenance, with a high impact on the user and on the purchase of his ticket.

In this perspective and taking into account the extensive analysis of the various systems and the possibility of obtaining real data and being able to demonstrate and validate the model based on the demonstration artifacts, it is intended to control the flow of data from the logs of this alarm system, ensuring the reliability of log production and sending between different entities over the internet, using blockchain technology to authenticate logs and control access to logs for entities, users and applications involved in this data flow.

In a process of iteration and improvement, it is intended in a second iteration to solve the problem of guaranteeing authenticity at the time of registration of the event and the way in which these records can be made available in the accepted nodes of this private network of different entities.

Further iterations would be necessary to achieve the entire data flow from these IoT devices and their availability in the smart city control boards, covering all five levels provided for in the smart city generic data model, in point II.2.1.

4th Stage - Demonstration

After the development, it is intended to demonstrate the monitoring and alarming component of the ticketing system, to verify if the predicted proof-of-concept artifacts can guarantee data security and control the data flow.

In the iterative process foreseen in the methodology, the artifact can be extensively adapted to improve the way we can guarantee the participating Organizations confidence in the proper functioning of the system and mainly to minimize the impact on the end users of the system, the user, the citizen.

5th Stage - Evaluation

The evaluation results from the demonstration and verification that the logs obtained and transmitted prove to be immune to tampering or manipulation and that delivery is guaranteed exclusively to authorize recipients.

This process provides for a comparison between systems designed without this security concern and those that arise from a design supported by security by design through the implementation of these mechanisms based on blockchain technology.

At this stage, based on the evaluation carried out, a second iteration is considered for the 3rd and 4th stages of this method, reinforcing the authenticity processes and mainly the integrity of the data and the respective records.

6th Stage - Communication

The communication process is based on the writing of manuscripts on three essential aspects for the elaboration of this work embodied in this thesis, the systematic review of the literature, the generic reliable data model of a smart city and the data markets. For this purpose and in the course of the research and investigation work, we prepared three manuscripts that were presented at the conferences of the World Conference on Information Systems and Technologies (WordCIST) of 2018 and 2019 and of the International Conference on Software Process Improvement (CIMPS) of 2018, respectively. with the titles "Systematic Review of Literature, Research on Blockchain Technology as Support for the Proposed Trust Model Applied to Smart Places", published March 28, 2018, "A blockchain-secure smart city model",(Brandão et al., 2018b, 2018a, 2019). These submitted manuscripts are referenced in the Publications section at the end of the thesis. An initial manuscript was also prepared on the application of blockchain technology in health, specifically in clinical records and which is still under review and in research and deepening work based on the model adapted to the one presented in this thesis.

The contribution of this research work is revealed in the generic data model supported by blockchain technology applied to a smart city. The proofs of concept developed are oriented towards validating the model and enhancing the use of blockchain technology in IoT environments in the alarm and monitoring components.

The main limitations are the fact that the alarming and monitoring components were mainly evaluated and not the operations or transactions. A part of the data flow between the 5 defined levels was also evaluated.

However, the evaluation carried out with the two iterations in the DSRM process methodology, allows the demonstration that the proposed model is validated with the proofs of concept carried out.

In a broader context, the model can see its application extended to other smart places, which support Organizations, with other architectures and applications.

1.3. Central Problem

Smart places are vulnerable to data compromise (Popescul & Radu, 2016) and false data injection (K. Zhang et al., 2017), which with the growing volume of data and the large number of devices, spaces, infrastructures and users connected, extends the risks and can cause the commitment of the entire system(s), with the use of weaknesses that can be transmitted or exploited between systems. In an extreme case, the exploitation of weaknesses or injection could compromise the functioning of the city itself and, at the limit, disconnect or even destroy the physical infrastructure to the point of threatening the lives of citizens or their well-being. (Popescul & Radu, 2016)

The question to be answered is the following: how to guarantee that the data contained in the logs of the alarm systems are not tampered with or manipulated before being delivered to other entities and systems?

The log records equipment alarm events to help detect and protect organizations against cybersecurity breaches and improve maintenance processes. A log provides error alarms, activation and deactivation actions of functions or equipment, audit records of network, application or database access, how they accessed and when they had access. In this way, it allows the availability of systems to be managed and unauthorized access to systems or confidential information to be detected, allowing for the investigation of unauthorized behavior.

2. Search results

As already mentioned in the introduction, the proposed model must translate the dynamics of digital ecosystems that support natural ecosystems, the smart ecosystems, which respectively comprise several application domains, with data models and database "powered" by IoT devices and furniture.

The work focused mainly on the control of the data flow, having been developed artifacts, based on proofs of concept that test the guarantee of authenticity and integrity of the data, controlling the flows between entities and between applications allowing the design of a system based on in blockchain, supported on the IPFS network and using a blockchain network to transfer the control information about the data and access to the log files, avoiding the manipulation of the logs and in a second iteration of the records, preventing a participant from being able to make changes/omissions of events, distorting information, compliance with certain SLAs or the business logic of the monitored systems.

The evolution of this system in the different phases of the data flow can also offer the chance to use this control and security mechanism, from sensors, IoT devices or mobile equipment, avoiding manipulations and changes that avoid detection and condition the data of the monitored systems. .

2.1. Choosing blockchain technology

The question of why blockchain technology was chosen as an answer to the research questions is central. This choice, although it soon revealed itself as the emerging technology with the potential to make the proposed model viable, its characteristics were analyzed and verified throughout the theoretical study and specifically in point 2., of theII Chapter.

As we will see, this choice mainly focuses on the Organizational motivations for adopting blockchain technology. Connecting to other more common questions. What is blockchain technology? How can blockchain technology be applied to different business contexts? What potential uses might they have?

(Li et al., 2018) refer that future studies should focus on why and who, at the same time, to assess the impacts of blockchain technology at different levels, no longer considering blockchain technology as a black box, analyzing it in the contexts in which it should be use and under what circumstances this technology works best and for whom.

Thus, at this point, we seek to present the comparative analysis with other technologies, what are their advantages and disadvantages, so that it can be said, in this research context, that blockchain technology may be the best option.

The comparisons of blockchain technology with other technologies are mainly revealed in two perspectives. From a database perspective and from a security technology perspective.

2.1.1. Database perspective

From the database perspective, it appears that it is a technology that stores data in a distributed and immutable way, with the advantages and disadvantages that these characteristics represent.

The Table III.2 compares blockchain technology, with the legacy centralized database and the distributed database, through the main characteristics of the databases, such as availability, record integrity, error tolerance, confidentiality, computation time and collaboration between reliable nodes, at three levels, high, medium and low.

Table III.2– Blockchain vs. Centralized legacy database and distributed database

characteristics	*Blockchain*	Centralized legacy database	distributed database
Availability	High	Low	Average
Registry Integrity	High	Average	Average
Fault tolerance	High	Low	High
confidentiality	Low	High	Average
computing time	Low	High	Average
Collaboration between trusted nodes	High	Low	Low

(Anwar, 2017)

As seen in this comparison, blockchain technology has a high error tolerance, eliminating the possibility of a single point of failure, such as a distributed database.

Blockchain technology has a differentiating feature such as high collaboration between trusted nodes. Each node in the network thus contains the same version of the transaction history as all other participating nodes. This feature leads to high availability and the integrity of the log of transactions distributed by the nodes. Each registration is processed and verified by additional nodes. Another important feature, already mentioned, is the high fault tolerance, since there are identical copies of the ledger, with exact and updated information. The distributed nature of the blockchain-based system allows nodes to recover lost transactions after failure.

The Table III.3 seeks to compare nodes between permissionless (public), permissioned (private) blockchains and a centralized database, through the characteristics of throughput, latency, number of readers, number of writers, number of untrusted writers, mechanism of consensus and centralized management.

Table III.3 – Compare nodes between permissionless, permissioned and centralized database blockchains.

Characteristics	*Blockchain* without permission	*Blockchain* with permission	central database
Transfer rate	Slow	High	Very high
Latency	Slow	Average	Fast
Number of readers	High	High	High
Number of writers	High	Low	High
Number of unreliable writers	High	Low	0
Consensus Mechanism	Mostly PoW, some PoS	BFT protocols (eg PBFT)	None
Centralized management	No	Yes	Yes

Adapted from table 1, from Wust & Gervais (2018)

The exchange of data in a decentralized system depends on the scale of the system, with the number of writers without mutual trust, their throughput, and the number of states they update and which they can handle in a given period of time. When making the decision to use blockchain technology or not, consideration should be given to balancing these characteristics and how to compensate for the most negative characteristics, and in the case of blockchain technology, the weighting between public and private blockchains. (Wust & Gervais, 2018)

2.1.2. Security perspective

The security perspective is broader and more emphatic, and leads mainly to comparison with cryptographic technologies and other distributed ledger technologies (DLT- Distributed Ledger Technology), through numerous application possibilities or application attempts.

The most commonly used approaches to ensure data security are through public key infrastructure (PKI - Public Key Infrastructure) and encryption protocols, for example e-mail such as S/MIME (Secure / Multipurpose Internet Mail Extensions). However, on these technologies there are several security threats, namely the MITM (Man-In-The-Middle) attack and the EFAIL attack (security breach in email systems where content can be transmitted in an encrypted form).

As a solution to the limitations of these technologies, in the case of messages, Khacef & Pujolle (2019) present a proposal for using blockchain technology to make communications more secure, in messages, maintaining the performance and security of data recorded in the blockchain, through a "smart" contract to verify identities and associated public keys, and thus validate user certificates. The system is decentralized and allows users to exchange messages securely. The immutability feature of blockchain technology provides the solution to the problems identified in the field of centralized PKI.

Trust and traceability are two of the basic features of blockchain technology. However, these features may not be enough to provide a complete solution, requiring strong cryptographic protocols to be added. This add-on provides more confidence, traceability, security and control, essential for critical solutions in critical and risky contexts. Data immutability and traceability are fundamental requirements for any critical system. The system based on blockchain technology must ensure: the integrity protection of secure storage; data privacy and ownership; data sharing; and data traceability and accountability. (Katuwal et al., 2018)

2.1.3. *Blockchain* - advantages and disadvantages

Given the growing dependence on technology, people, companies and governments base their actions and decisions on information systems whose data they have to trust. Trust in digital spaces leads to a relationship with trust in "authorized" data. Thus, breaches of trust, forgeries install doubts and fears, with new security and privacy problems, and with the spread of dematerialized transactions. (Ruta et al., 2017)

The Table III.4 presents the main strengths and weaknesses of blockchain technology in relation to alternatives and in certain cases.

Table III.4 - Strengths and weaknesses of blockchain technology

Strong points	weaknesses
Visibility	Lack of privacy
Aggregation	Lack of standardization
Validation	*Garbage in, garbage out (GIGO)*
Automation	Black box effect
Resilience	Inefficiency

Adapted from table 1, from Babich & Hilary (2019)

The strengths of blockchain technology and its weaknesses reflect the need to consider the objectives of the systems to be developed or replaced.

The main advantage of blockchain technology is that the ledger cannot be modified or deleted after the data has been approved by the nodes as per the consensus protocol. These characteristics allow Blockchain technology to guarantee data integrity and security, which is why its use also extends to other services and applications. (Rithika et al., 2019)

Blockchain technology thus provides a platform that allows the interconnection of various entities, with multiple sources of data and that generate information to support decisions. Thus avoiding that the decision can be conditioned or stopped with erratic information or used for unintentional purposes or in false records. (Babich & Hilary, 2019)

One of the main problems with other technologies is how to verify that the information received from the network is authentic and updated. Blockchain technology has the ability to solve this authenticity problem without including trusted intermediaries. This feature allows any stakeholder to independently verify the authenticity of the data, from who or where the data is. (Mattila, 2016)

The main attributes of blockchain technology are eliminating the need for intermediary trust authorities, due to the decentralized nature of the network, the increased transparency and immutability of the blockchain. (Hammer, 2018)

According Lin & Liao (2017) blockchain technology is comprised of six key elements: decentralized; transparent; open code; autonomy; immutable; and anonymity.

2.1.4. Blockchain in the IoT Context

Issues related to smart cities, which are based on networks with IoT devices, go through bottlenecks, with the multiplier increase in connected devices that create problems in a centralized system. Blockchain technology can be applied to solve this problem due to its decentralized feature. Amjad & Javaid, (2019) presented a work that proposes a hybrid network architecture with centralized and other decentralized functions, to obtain efficiency. The blockchain-based system allows securing the devices from untrusted services provided by untrusted servers. Blockchain technology is intended to keep the system secure for different transactions whenever services are provided to the customer. The client is protected, as the services are verified at the beginning. The analysis shows that the system performance increases and security increases against the existing system.

The IoT environment needs a registration and authentication process, in which blockchain technology, in relation to cryptography, can provide better forms of authentication and in the registration process than with other contemporary solutions. (Ghuli et al., 2017)

Blockchain technology requires that most nodes on the network must participate in the consensus mechanism to verify the source data. IoT devices are devices with limited resources, so it is necessary to improve the existing blockchain scheme to allow these devices to be included. Chen, Wang, and Wang (2018) present two possibilities to overcome this possible limitation, through: random selection of cooperative nodes to solve the hash puzzle to reach a consensus; and majority-based verification without needing to perform encryption (encrypt/decrypt) reducing the computation of each IoT node.

The potential of blockchain technology to protect data integrity in IoT networks can be enhanced through blockchain-based data verification scheme and overcome the shortcoming of centralized approaches with single point of failure and network congestion. This proposal can reduce the number of cooperation nodes and rely on nodes to generate the block, and the data is relayed by a random number of randomly selected cooperation nodes. Thus, the security level of the system can be significantly improved.

The Table III.5 summarizes the comparison between PKI and blockchain-based data scheme with verification scheme to prevent central node attack, eliminate network congestion and prevent link attack.

Table III.5 - Comparison between PKI and stochastic blockchain based data schema with verification scheme

	PKI	**Blockchain based data with verification scheme**
Avoid central node attack	Certification may fail when Certification Authority (CA) is destroyed.	Certification cannot be affected by a single point attack.
Eliminate network congestion	Only the CA has the authority to verify.	Each node has the authority to verify.
Avoid the attack of links.	Attacker can attack access to Certificate Authority (CA) at source and destination	The attacker does not know who to attack.

Adapted from figure 8, from Chen, Wang, and Wang (2018)

The simulation results demonstrate that security has increased in a large-scale IoT network, even with a small number of cooperative nodes.

Cekerevac, Prigoda, and Maletic (2018) refer that blockchain technology can identify and authenticate standalone IoT devices. In this way, blockchain technology can transform IoT solutions, avoiding failures in recording data and information, allowing to control sensors and measurement data, and IoT equipment to be able to send the data directly. The technology allows each participant, based on the license level, to access data, namely the status in real time.

2.1.5. Adoption

Non-technical drivers, namely "philosophical beliefs", network effects and economic incentives, could also explain the adoption of blockchain technology. These drivers may explain, after all, the underlying reason for adopting blockchain technology.

Table III.6 - Drivers for the adoption of Blockchain technology

Category	Drivers
Scenario properties	Storage status
	Various writers
	Cannot use TTP (Trusted Third Party)
	Unknown writers
	Untrustworthy writers
	Public verification
"Philosophical Beliefs"	TTP will not be used
	Need for decentralization
	Privacy enhancement
	Alternative system
	Political reasons
Network effects	Community driven
	Curiosity
	Good to use
Economic incentives	Marketing product
	Mining sales equipment
	Sales consultancy
	Upload to platform
	Fear to lose
	Alternative investment

Adaptation of table 1, from Koens & Poll (2018)

In addition to these drivers, it is necessary to be more objective and seek answers and results that allow the sustainable adoption of this technology.

What started as a code placed by an anonymous researcher (Nakamoto, 2008), with the aim of creating a new currency platform, bitcoin, blockchain technology has skyrocketed in popularity, in almost all sectors, from supply chains to finance, and health, through education and city planning. In conclusion, blockchain technology seems to improve not only tasks in current industries, but also in other emerging contexts. (Sharma & Bhuriya, 2019).

As far as blockchain capabilities like increasing data access and data validity are concerned, blockchain technology appears mainly with the security aspects, with confidence in the validity of data, where it comes from and for tracking actions. (Bauer, Zavolokina, Leisibach, and Schwabe, 2019).

In order to test the adoption of blockchain technology in the proposed model, the maturity model for blockchain adoption proposed by Wang (2016), which is described below, with the three-phase procedure for the safe adoption of this technology.

Point 1: Why Blockchain? (4 out of 6 positive questions allows for safe adoption)

- Multiple parties share data: multiple stakeholders need to see common information;
  Does it apply to our proposal? Yes. To various entities and/or applications.
- Multiple parties share updated data: multiple participants take actions that need to be recorded and change the data;
  Does it apply to our proposal? Yes.
- The verification requirement: Participants need to trust the validity of the actions that are recorded;
  Does it apply to our proposal? Yes.
- Intermediaries add cost and complexity: elimination of intermediaries and the "central authority" of registration has the potential to reduce cost and complexity;
  Does it apply to our proposal? Yes.
- Interactions are time sensitive: reducing delays has business benefits;
  Does it apply to our proposal? Yes.
- Transaction interaction: transactions created by different participants depend on each other;
  Does it apply to our proposal? Yes.

Result: The answers to 6 of the 6 questions are positive. It is a safe adoption according to the maturity model for blockchain adoption.

To reinforce its adoption, points 2 and 3 must be respected, which is in line with the proposed model.

Point 2: Development should focus on:

- Requirements analysis;
- *design* architectural.

Point 3: Operation, in which the blockchain system will replace an existing system, proposing a progressive replacement procedure:

- Keep the existing system working and run the blockchain system as the backup system for a certain period;
- If the blockchain system is working properly, let it run as the main operating system and run the existing system as the backup system.
- Finally, operate the blockchain system as the support.

Result: Our model and the developed artifacts start progressively, seeking to complement the entire data flow.

As evaluated, the adoption of blockchain technology provided for in the generic data model, in the context of smart places, supported by IoT, meets the requirements that can enhance the advantages listed throughout this point, mainly in ensuring data authenticity and integrity. of data, providing several applications that will be complementary and structuring in the functioning of smart city ecosystems.

2.2. Generic Data Model

The result of the theoretical study carried out allows the structuring of general ontologies that prove to be potentiators of a generic data model that represents the set of concepts and the relationship between them, inferring the main objects of the domain.

The proposal of a generic model of trust data represented in the Figure III.2 intends to organize a global smart city, in ecosystems (Ceballos & Larios, 2016; Dhungana et al., 2016), with their respective application domains, supported by generic data models, fed by numerous IoT data.

The smart place is the physical and logical space that is intended to be structured. Smart cities are divided into ecosystems, smart economy, smart environment, smart living, smart mobility, smart people and smart government, based on the circle of Cohen

(2013). Each ecosystem has several application domains with management, monitoring, security and privacy components, which are structured in generic data models, depending on the domain's applications.

The model has five stratified levels, which aggregate the data and summarize at the top level the necessary and sufficient information to guarantee the management of the lower level.

This model reflects the design, as the synthesis of the study (Walls et al., 1992) carried out in order to simplify and reflect the perspective of a smart place (smart city) based on IoT.

Figure III.2 Context model of a smart city

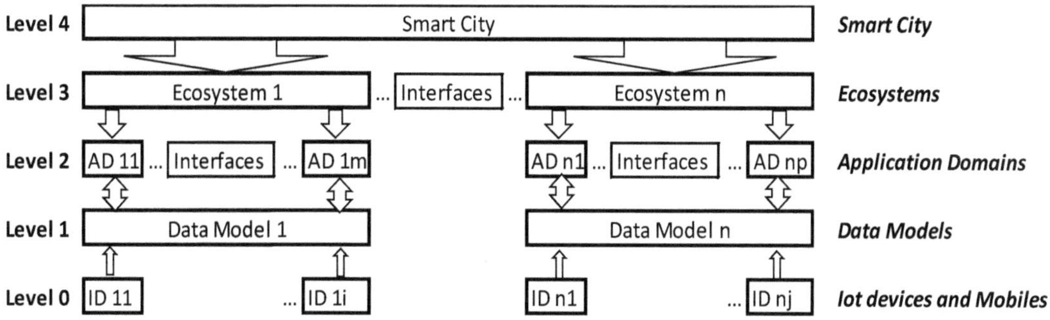

Level 4	Smart City	*Smart City*
Level 3	Ecosystem 1 ... Interfaces ... Ecosystem n	*Ecosystems*
Level 2	AD 11 ... Interfaces ... AD 1m AD n1 ... Interfaces ... AD np	*Application Domains*
Level 1	Data Model 1 Data Model n	*Data Models*
Level 0	ID 11 ... ID 1i ID n1 ... ID nj	*Iot devices and Mobiles*

(Brandão et al., 2018a)

In this model, the exchange of data between ecosystems is based on interfaces that promote the interoperability of data that need or depend on data from other ecosystems. This is a fundamental aspect to manage data and control the flow of data and to achieve unique and reliable data for the dynamics expected of a smart city.

The characteristics of the data must allow the flow of data to be consistent and sufficiently granular at each level to be able to segment shared data with the appropriate protections and permissions.

Each tier has data processing and storage capacity, through distributed cloud computing, IoT data reuse and aggregated data.

The intended model of trust is oriented towards making the system centered on the citizen, on the user, with a global approach to data management and data governance, with the control of permissions, consents regarding privacy, visualization of data, access to histories and the reuse of data for different purposes.

THE Table III.7 intends to group the set of blockchain applications in a smart place (in the smart city), typifying into six types of blockchain application that group the purposes that support the model.

In a first type, we have secure transactions that are oriented towards the use of blockchain technology to guarantee distributed transactions of immutable data, with the exchange of consistent and incorruptible transaction records and, if possible, oriented towards the monetization of transactions.

In a second type, we have data security that allows the use of blockchain technology to guarantee authorized access to the various participants or involved in the processes, complying with standard or authorized permissions, in a manner consistent with the different types of actors and by the action of the citizen or user. .

In a third type, we have the control of the data flow that allows the use of blockchain to guarantee the management of data between Organizations and respect for the rules associated with the types of data (personal, sensitive personnel, sensitive data, open data, etc.). This use also allows citizens to access their data, consent and access history to their data.

In a fourth type, we have the acceptance of devices that guides the use of the blockchain to ensure that the accepted nodes have an "identity" in the Organization, validated by certification or through smart contracts.

In a fifth type, we have version control that guides the use of the blockchain to ensure the compatibility of versions of the system, IoT equipment, mobile devices and other devices.

In a sixth type, we have the application in the security of the system that intends to use blockchain technology to ensure the trust of system configurations, such as cloud servers, IoT devices and mobile devices.

The following table shows these six types of blockchain technology application.

Table III.7 - Blockchain Application Types (BC)

Type	Blockchain Applications (BC)
BC1	secure transactions
BC2	data security
BC3	Data flow control
BC4	Device acceptance
BC5	version control

Type	Blockchain Applications (BC)
BC6	Systems security

The Table III.8 lists the five levels proposed in the Figure III.2 with the six types of application of blockchain technology from Table III.7.

Table III.8 - Application of blockchain in the context of smart cities

smart city context		Blockchain application					
		BC1	BC2	BC3	BC4	BC5	BC6
Level 4	*smart city*	X	X	X			
Level 3	ecosystems	X	X	X			
Level 2	Application Domain	X	X	X	X	X	X
Level 1	data models		X	X		X	X
level 0	IoT and mobile devices		X	X	X	X	X

Analyzing what is intended at each of the levels, we found that the BC2 type of data security and BC3 of data flow control apply to all levels in the context of the smart city. BC1 type of secure transactions applies to the top three levels. Types BC5 and BC6, referring to version control and system security, apply to the 3 lower levels. BC4 type applies to level 2 and 3, application domain and IoT and mobile devices.

2.3. General Architecture

The definition of the architecture is based on the interaction flows of the different levels and types of blockchain application revealed in theTable III.8using blockchain platforms and IoT platforms. Data flow control is based on the need to segment data into groups with homogeneous characteristics and that allow their management and control of data flows, as a way of defining security and privacy rules that allow controlling access and amendment, subject to permissions management and consent management. The Figure III.3 presents the main flows in a smart place, respecting the six applications of blockchain technology provided for in the model.

Figure III.3 - Data flows in a smart place

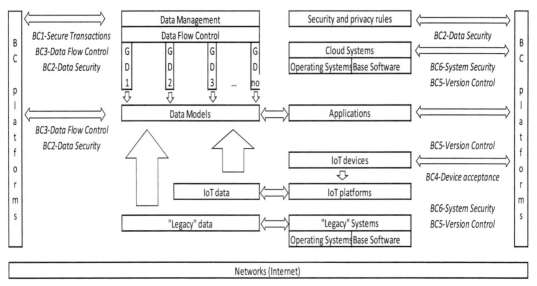

The proposed trust model presents as a critical aspect the way to establish interoperability based on data groups (DG). Data groups respect data structures, according to the application domain, to allow defining granularity over the dataset, to apply privacy rules and to apply standard security rules or those defined by the citizen/user.

Smart places focused on the citizen/user and from a perspective of physical and logical mobility will have to interact with each other and communicate the data groups and the security and privacy rules that will be available to establish a continuity of data and citizen/user presence. in different physical and logical spaces.

This perspective can be seen in Figure III.4 through macro flows intra and inter smart places, centered on the citizen/user, which communicate within their space (intra) and between distinct spaces (inter) through data groups and in accordance with standard or defined security and privacy rules by citizens/users.

Figure III.4 – Macro Flows Intra and Inter smart places, citizen/user centered

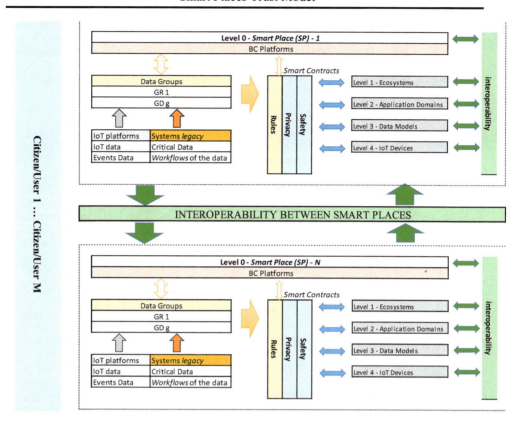

The relationship that may be established between various smart places will not be analyzed in this work and the application of the defined methodology, and may be implemented in future investigations.

2.4. Application of the methodology

At this point, the summarily descriptive methodology is developed in the point of this chapter, specifically in point III.1.2. regarding the steps of the methodology.

The implementation of the methodology will allow the validation of some aspects of the generic data model based on blockchain technology and the general architecture described in the previous points.

This specification leads to a clear and objective proposal for the realization of artifacts that will allow to demonstrate and validate the model and its projection or extrapolation to the remaining ecosystems, application domains, systems and data models and IoT devices that constitute the generic data model proposed for smart places and in this case under analysis in smart cities.

2.5. The choice of the mobility ecosystem and the application of Ticketing

In the context of smart cities, issues of authenticity and data integrity are critical in the functioning and security of the various ecosystems and people, as detailed in section 1.1., of the chapter II.

The research team's in-depth knowledge of this ecosystem and the possibility of intervening, developing and reviewing, in terms of data, database and applications, allowed the consolidation of several aspects to be evaluated in the proposed model.

The mobility system, analyzed in point4, from the chapterII, in the rail transport component, in addition to railway safety concerns, additional regulatory aspects, integration with other modes of collective and individual transport, imperative needs for availability, compliance with the offer, quality of services provided to the customer and multiple access to monitoring and alarming that allow in real time the redefinition and reallocation of transport, self-regulation, anticipation of problems and safety in all components.

The initial choice focused on the traffic management system in its alarm and monitoring component, as a system of high criticality and risk, with an impact on the safety of people and public space. In the studies carried out, it was possible to verify that it was a system subject to major access restrictions, limitations of intellectual property, patents and proprietary software, which conditioned the development of the artifact, the knowledge of the application's data model and the possibility of validating and test. Although it was possible to test the authenticity of the logs referring to the 1st iteration, demonstrating the universality of the model and the proposed artifact.

The ticketing system emerged as an option to validate the model as an answer to the research question. This system has a high impact on the end customer, on their perception of the quality of the mobility system, economic sustainability and environmental impact in spaces and smart cities. The monitoring and alarming component provides data on the multiple sales, validation and sensing equipment, distributed by various entities/operators, with multiple providers subject to service levels based on high availability of equipment, especially those with a high direct impact. in the final customer and in the most affluent locations.

As will be verified, the chosen system will serve as a reference for the realization of the artifact, with the proofs of concept, and can be extrapolated to other systems that are part of a smart place, in the alarmistic and monitoring components.

The in-depth knowledge of these systems in the mobility ecosystem allowed us to build proofs of concept and validate the proposed generic data model.

2.6. Artifact design and development

The design and development of the artifact was supported by the choices supported by the points 2.1 and 2.5 that allowed finding a way to validate the proposed model, through the use of blockchain technology in the context of the mobility ecosystem.

The steps described in 1.2 and 1.3, of chapter 3, define the problem and motivation.

Smart places are vulnerable to data compromise (Popescul & Radu, 2016) and false data injection (K. Zhang et al., 2017), using weaknesses that can be transmitted or exploited between systems.

The design and development of the artifact aims to find the solution that allows testing the application of the reliable generic data model based on blockchain technology, supporting smart places, particularly smart cities.

The artifact thus intends to test and validate the proposed model in a specific component, to guarantee data security and data flow control.

The data model and the data that serve as the basis for the artifact refer to the application component of monitoring and alarming events, existing in IoT-supported applications and to which the proposed model is directed.

The recording and treatment of these events allow for continued actions of predictive, preventive and corrective maintenance, the assessment of the availability of systems and equipment and the calculation of the service levels contracted for support and maintenance.

The smart place proposed for this experiment is a smart city, the ecosystem under analysis is the mobility ecosystem (smart mobility), the application domain is urban public rail transport, the system and data model that serve as the basis for the methodology is that of the ticketing system in its alarm and monitoring component of IoT equipment.

Compliance with service levels, through contractual Service Level Agreements (SLAs), has a high impact on the end user on the purchase of their ticket, with consequences for possible fines and satisfaction or dissatisfaction with the experience of traveling on a particular transport.

The manipulation of data can hide existing problems and hide serious situations of unavailability, improper access and serious or critical alarms that can be omitted, with unforeseeable consequences.

The choice of this system and the data model to test the model was based on the high impact it has on the final customer and on the perceived quality it has in shaping the overall quality of the transport system. Also because it is an application component transversal to the systems and applications that are supported in equipment with sensors or based on IoT and in this way to be able to demonstrate and validate the model in this alarmistic and monitoring component based on the proofs of concept that make up the artifact of demonstration.

Other systems were analyzed and considered for this work, mainly the traffic management system, which proved to be very difficult to access given the level of security and the security certification processes it complies with and the lack of access to the data structure and model, highly conditioned by the fact that it is a proprietary system and highly dependent on the supplier and with restrictions on intellectual property, copyright and patents.

This artifact evolves from the traditional model supported on a log file that is transferred between entities or applications, to the trust model of effective control of the data flow of the logs of the alarm system (in the first iteration of the methodology) and in the security in the registration of the data (in the second iteration of the methodology).

2.6.1. 1st Iteration of the proposed methodology

The artifact, based on proofs of concept, in the 1st iteration, intends to guarantee trust in the generation of log files, in their access and in sending them between different systems or entities.

The traditional global data flow presents the data flow of alarming and monitoring events to the central database and then its availability for monitoring applications, for the creation of sequential log files in time and their transfer to other entities and for applications that consolidate this data in applications such as SIEM (Security Information and Event Management), datawarehouse, SLA calculation tools with penalties and BI (Business Intelligence) tools.

Figure III.5 - Data flow of alarming and monitoring events

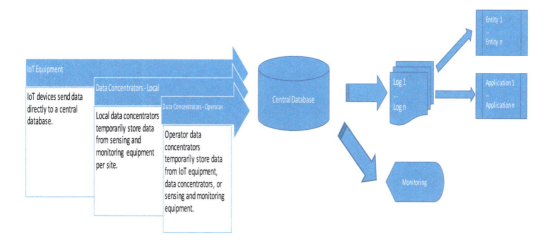

The existence of local and operator data concentrators is due to the need to store data in offline situations in which various equipment does not have its own storage or has limited capacity.

The design of the artifact focuses on the stage indicated in the following figure.

Figure III.6 - Artifact object data flow phase

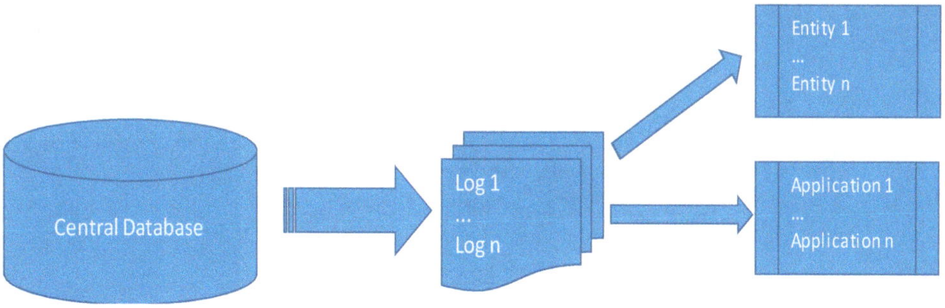

This first iteration focused on answering how to protect logs and ensure their authenticity. The log records events and alarms, allowing organizations to analyze and protect against cybersecurity breaches. Logs can provide audit records of network, application or database accesses, how they accessed and when they had access. Thus, they allow the detection of unauthorized access to systems or confidential information and the investigation of unauthorized behavior.

The main problems are security breaches in accounts with special privileges or permissions, to obtain a level of access that allows editing the logs to manipulate their trace and the historical log.

The traditional flow of Figure III.6 has been changed to allow for two processes described in Figure III.7 flow control, which pass through:

- The generation of log files and making them available on an IPFS network supported by blockchain, providing the various nodes of the private network with access by private and public key for the encryption/decryption of the available files.
- The data with the indexing of the file links and respective hash, distributed by a blockchain network, provide the nodes of the entities involved with the information to access the files, in sequence and with the hash expected for each of the files (the same can be used by the IPFS network access hash).

THE Figure III.7 thus presents the design of the demonstration artifact, ensuring the redundancy of log files intrinsic to the IPFS network and access data on the private blockchain network, the authenticity of log files through file hash mechanisms and access supported by keys for encryption/ file decryption.

Figure III.7 - Artifact Design in the 1st Iteration

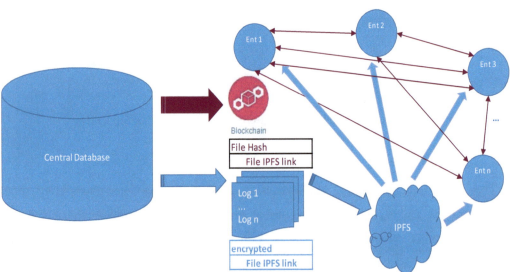

The private blockchain network requires an invitation and must be validated by the entity or user that implemented the network, complying with the set of implemented rules. This network configures a private blockchain and an authorized network that imposes restrictions on participation and transactions to be carried out. In this way, participants need permission to participate, through the access control mechanism using a certification authority or with the issuance of private and public network keys.

The private and independent blockchain implemented in this proof of concept allows the integration process with other entities or applications to provide an independent verification of the authenticity of the log files, the integrity of the data in the file and the sequence of the files.

The encrypted IPFS file is accessed through its hash (ipfs/ <filehash>) contained in the blockchain in order to decrypt it using the private key. Key management and distributed blockchain sharing allows managing access to log files with IoT data. The implementation of the encryption mechanisms and the sending of log files to IPFS took place in accordance with the details described in point2.6.1.1 with the .

The blockchain network integrated with IPFS through the Infura platform was also analyzed.[63]which provides the APIs to access Ethereum and IPFS, avoiding infrastructure-specific configurations for Ethereum nodes and IPFS nodes. This solution allows you to interact with Ethereum and IPFS platforms through the tools and APIs available.

The creation of this log file replication mechanism is maintained by the ledgers distributed by the different nodes of the private blockchain network, of different entities, with security properties, namely data immutability to access the log files accessible on the IPFS distributed network.

The 5th Stage, evaluating the first iteration of the investigation methodology followed, made it possible to verify that the implemented mechanisms allow guaranteeing the objectives defined in the design of the artifact.

---

[63] Infuriates - https://infura.io/, accessed on 06-10-2019

The evaluation, which resulted from the demonstration and verification, indicates that the logs obtained and transmitted prove to be immune to tampering or manipulation and it is guaranteed that they are delivered exclusively to authorized recipients.

This process compared traditional systems designed without this concern for authenticity and security, with this proof of concept designed and it was found that through the implementation of these mechanisms based on blockchain technology, the creation of solutions designed with a basis for security and privacy (security by design and privacy by design).

The evaluation confirmed that the use of blockchain technology can make access to the decentralized log available at any node on the network, with hash data available in these authorized locations and accessed through cryptographic mechanisms that protect them from unauthorized access.

Nodes can validate by consensus and verify that the data is correct, making it impossible to edit the log and manipulate it. Each block in the blockchain has the unique hash of each log file, which proves the integrity of the information and the reliability of all transactions, preventing any attempt to tamper with the log and its unauthorized access by cryptographic protection.

The evaluation identified an important weakness in the control of data flow. The data sent to the log files and inserted into the database can, at this stage of insertion, condition the generation of log files and can be manipulated by the entity that holds the central database or by improper access.

In this evaluation and taking into account the flow of data, there was also a need to control the stage of the process prior to the generation of log files, with the risk that at this stage it is possible to filter or manipulate the records inserted in the database. This finding implied a 2nd iteration that proves to be important to complete another phase in the control of the global data flow.

2.6.1.1. Operationalization of the 1st Iteration

The 1st iteration was carried out with the implementation of the encryption mechanisms and sending the log files to the IPFS network and the implementation of the blockchain in Ethereum, and was carried out with the steps described and developed for these proofs of concept.

The Ethereum platform was adopted for the blockchain network, creating three nodes and developing the data structure, namely the hash of the log files and the links to the files on the IPFS network. In this way, the nodes of this private network contain the private and public keys that allow access to decrypt the log files.

In the first set of tests, referring to the 1st iteration, the system was analyzed in a context of normal operation although in a simulated environment.

In this case, and in the first iteration, real data collected until 2019-04-01 at 15:46:21 by the various ticketing network equipment were used, inserted in a test database, in SQLEXPRESS version 14.0.1., based on a daily export process where log files were generated with the events filtered by the type of equipment and with the types of alarms to assess compliance with availability service levels (SLA's).

The following table shows the fields that are recorded in the log file on a daily basis and that are sent to third-party service providers and to specific applications for calculating availability and respective application of penalties depending on the contractual indicator range.

Table III.9 - Fields exported to the Log file

Field	Description
Code	Event ID
Equipment	ID Equipment
EventType	Kind of event
DescriptionTypeEvent	Description Type Event
Module	Module
DateTimeOn	Date Time Online Actual
DateTimeOnCorrected	Date Time online Corrected to the monthly period
DateTimeOff	Date Time Offline Actual
DateTimeOffCorrected	Date Time Offline Corrected to monthly period
Station	Station Code
NameStation	Station name
Operator	Transport Operator
OperatorName	Operator Name

The following table identifies the types of events considered for this assessment. It should be noted that the types of events are mainly focused on those that have the greatest impact on the customer, in the use of the ticketing system.

Table III.10 - Type of Events considered

Event Type	Event Type Description
124	Full Coin Vault - Coin Vault
123	Coin Vault is not present - Coin Vault
132	Banknote vault full - Banknote vault
131	Note Vault is not present - Note Vault
164	Card dispenser blocked - Dispenser

Event Type	Event Type Description
224	Blocked card dispenser - Dispenser 2
121	No change ("exact amount" mode) - Change deposit
126	Stopped due to jam - Change deposit
160	Jammed card dispenser - Dispenser
220	Jammed card dispenser - Dispenser 2
161	Empty card dispenser - Dispenser
170	Communication Error (Receipt Printer) - Receipt Printer
171	Paperless - Receipt Printer
120	Communication Error (Coins) - Coins
122	Error opening shutter - Coins
134	Notes stuck in the mechanism - Notes
130	Communication failure - Notes
140	No communication with OEM module - OEM/SIBS
143	No communication with SIBS - OEM/SIBS
142	MB payment not available - OEM/SIBS
190	Communication Error (Contactless Module) - Non-Contact
102	Out of service

The perception of the overall functioning of the system is reflected in the calculation of indicators and should lead to more accurate replacement, conservation and maintenance processes that minimize their high impact on business processes. These are processes that require trust in the data contained in the log files, so it is tempting to manipulate the log files in order to omit or correct certain signals or their entries in ON or OFF. The periods of unavailability may be weighted more heavily (2x more) depending on the location of the equipment in view of the impact on the customer and the simultaneity of the unavailability of equipment at a location.

The types of events are treated according to the simultaneity of the signals, and may not count if they occur in the same period.

THE Table III.11 presents an example of the generated log file, referring to the month of November 2018 (from 11-01-2018 to 11-30-2018).

Table III.11 – Example of the log file

Codigo	Equipamento	TipoEvento	DescricaoTipoEvento	Modulo	DataHoraOn	DataHoraOnCorrigida	DataHoraOff	DataHoraOffCorrigida	Estacao	NomeEstacao	Operador	NomeOperador
49721319	1717	130	Falha de comunicação	Notas	2018-05-26 14:25	2018-11-01 00:00		2018-11-30 23:59	17	Casa da Música	2	Operador01
52479956	24912	123	Cofre de moedas não está presente	Cofre de moedas	2018-10-24 08:38	2018-11-01 00:00	2018-11-03 12:53	2018-11-03 12:53	249	Fânzeres	2	Operador01
52513980	2812	123	Cofre de moedas não está presente	Cofre de moedas	2018-10-26 12:41	2018-11-01 00:00	2018-11-02 14:51	2018-11-02 14:51	28	Vilar do Pinheiro	2	Operador01
52600364	1212	102	Fora de serviço		2018-10-31 14:25	2018-11-01 00:00	2018-11-01 09:41	2018-11-01 09:41	12	Srª da Hora	2	Operador01
52602213	5813	120	Erro de comunicação (Moedas)	Moedas	2018-10-31 17:12	2018-11-01 00:00	2018-11-01 14:58	2018-11-01 14:58	58	João de Deus	2	Operador01
52602681	24411	130	Falha de comunicação	Notas	2018-10-31 18:13	2018-11-01 00:00	2018-11-01 00:17	2018-11-01 00:17	244	Rio Tinto	2	Operador01
52602737	6111	120	Erro de comunicação (Moedas)	Moedas	2018-10-31 18:14	2018-11-01 00:00	2018-11-01 05:59	2018-11-01 05:59	61	24 de Agosto	2	Operador01
52603312	25015	131	Cofre de notas não está presente	Cofre de notas	2018-10-31 18:59	2018-11-01 00:00	2018-11-01 04:33	2018-11-01 04:33	250	Santo Ovídio	2	Operador01
52603619	5813	131	Cofre de notas não está presente	Cofre de notas	2018-10-31 19:23	2018-11-01 00:00	2018-11-01 16:28	2018-11-01 16:28	58	João de Deus	2	Operador01
52603626	5813	160	Dispensador de cartões encravado	Dispensador	2018-10-31 19:23	2018-11-01 00:00	2018-11-01 14:59	2018-11-01 14:59	58	João de Deus	2	Operador01
52603657	5813	140	Não há comunicação com o módulo OEM	OEM/SIBS	2018-10-31 19:27	2018-11-01 00:00	2018-11-01 14:55	2018-11-01 14:55	58	João de Deus	2	Operador01
52604256	1713	122	Erro na abertura do shutter	Moedas	2018-10-31 20:31	2018-11-01 00:00	2018-11-01 04:21	2018-11-01 04:21	17	Casa da Música	2	Operador01
52604416	1611	161	Dispensador de cartões vazio	Dispensador	2018-10-31 20:52	2018-11-01 00:00	2018-11-01 00:02	2018-11-01 00:02	16	Francos	2	Operador01
52604498	3412	130	Falha de comunicação	Notas	2018-10-31 21:04	2018-11-01 00:00	2018-11-01 03:20	2018-11-01 03:20	34	Fonte do Cuco - Linha P	2	Operador01
52604784	5412	140	Não há comunicação com o módulo OEM	OEM/SIBS	2018-10-31 21:36	2018-11-01 00:00	2018-11-01 08:45	2018-11-01 08:45	54	Aliados	2	Operador01
52605093	24412	140	Não há comunicação com o módulo OEM	OEM/SIBS	2018-10-31 22:19	2018-11-01 00:00	2018-11-01 03:38	2018-11-01 03:38	244	Rio Tinto	2	Operador01
52605136	5011	171	Sem papel	Impressora de recibos	2018-10-31 22:22	2018-11-01 00:00	2018-11-01 00:11	2018-11-01 00:11	50	Salgueiros	2	Operador01
52605250	2611	140	Não há comunicação com o módulo OEM	OEM/SIBS	2018-10-31 22:41	2018-11-01 00:00	2018-11-01 03:06	2018-11-01 03:06	26	Modivas Centro	2	Operador01
52605265	24612	120	Erro de comunicação (Moedas)	Moedas	2018-10-31 22:44	2018-11-01 00:00	2018-11-01 03:48	2018-11-01 03:48	246	Baguim	2	Operador01
52605469	312	130	Falha de comunicação	Notas	2018-10-31 23:13	2018-11-01 00:00	2018-11-01 10:42	2018-11-01 10:42	3	Brito Capelo	2	Operador01
52605655	6816	170	Erro de comunicação (Impressora de recibos)	Impressora de recibos	2018-10-31 23:37	2018-11-01 00:00	2018-11-01 08:54	2018-11-01 08:54	64	Estádio do Dragão	2	Operador01
52605794	812	134	Notas encravadas no mecanismo	Notas	2018-10-31 23:51	2018-11-01 00:00	2018-11-01 11:04	2018-11-01 11:04	8	H. Pedro Hispano	2	Operador01
52605983	5011	140	Não há comunicação com o módulo OEM	OEM/SIBS	2018-10-31 23:58	2018-11-01 00:00	2018-11-01 00:11	2018-11-01 00:11	50	Salgueiros	2	Operador01

The introduction of a control layer, proposed in the 1st iteration, results from the need to generate, save, send and receive log files, from and to different Entities (depending on the greater or lesser division of maintenance services) or calculation or of predictive and preventive analysis, which seeks to ensure confidence in the data and improve the overall functioning of the ticketing system.

The test environment was created using Oracle VM Virtual Box version 5.2.26, where nodes were created with the Ubuntu Server 16.04.6 operating system (BC01, BC02 and BC03)[64], installing the gpg and ipfs tools.

The proofs of concept carried out below made it possible to generate the keys and encryption mechanisms, providing the necessary data for the nodes of the blockchain network, for accessing log files and the way to decrypt authentic log files.

The following commands made it possible to generate the key pair, the private key that must be kept safe and the public key that can be shared with other users who can be trusted with access.

On node BC01:

- *BC01>gpg --gen-key (RSA 2048, with username: dwbctest01, email: dwbctest01@gmail.com)*

---

[64] https://itsfoss.com/install-linux-in-virtualbox/, updated and accessed on 2019-08-28

To export the public key of dwbctest01:

- *BC01>gpg --export --armor dwbctest> dwbctest01PublicKey.asc*

To import the public key on other nodes (BC02 and BC03) with other users (dwbctest02 and dwbctest03):

On node BC02:

- *BC02>gpg --import dwbctest01PublicKey.asc*
- *BC02>gpg --list-keys*

On node BC03:

- *BC03>gpg --import dwbctest01PublicKey.asc*
- *BC03>gpg --list-keys*

To encrypt the log file:

- *BC01>gpg --encrypt - dwbctest02 DWBC_BI_20181203111116_20181101_20181130.csv*

To send the log file to IPFS:

- *BC01>ipfs add DWBC_BI_20181203111116_20181101_20181130.csv.gpg*

The expression Qm ...1 is the hash of the original log file.

Users dwbctest02 and dwbctest02 get the file through the following commands:

On node BC02:

- *BC02>ipfs get Qm...1*
- *BC02> gpg --decrypt Qm...1 > DWBC_BI_20181203111116_20181101_20181130.csv*

On node BC03:

- *BC03>ipfs get Qm...1*
- *BC02> gpg --decrypt Qm...1 > DWBC_BI_20181203111116_20181101_20181130.csv*

THEFigure III.8 presents how the blockchain associated with IPFS only stores the hash of the IPFS file (and/or its link), keeping the data required on the blockchain in its simplest form, with IPFS distributed peer-to-peer properties.

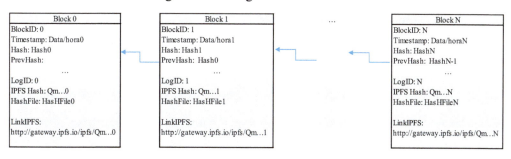

Figure III.8 - Log control blockchain

The fields IPFSHash (SHA256), HashFile (if necessary a stronger hash like SHA512) and LinkIPFS, added to the base structure of the block, allow the blockchain distributed by the different nodes of the private blockchain network to contain the necessary and sufficient information to guarantee the authenticity of the log files, the possibility of verifying their integrity and sequence, and through access to the public key on the other nodes (BC01, BC02 and BC03) for the possibility of decryption.

The implementation of this private blockchain network is based on 3 nodes and supported on the Ethereum platform as per the procedures[65],[66],[67],[68] and [69] adapted.

The joint configuration of IPFS and Ethereum can be performed through Infura according to procedures [70] adapted, in which log files can be stored on IPFS and the hash of the log file can be stored on Ethereum, through the use of APIs to access the Ethereum network and IPFS.

It was also possible to perform the 1st iteration tests for the logs generated by the traffic management system, through the generated log files (TT_LOG - 2018-12-02-0520.csv and Event_LOG - 19010100.csv), referring to the passages of the vehicles at fixed beacons along the entire track, thus verifying the possibility of guaranteeing the authenticity and integrity of the data, through a layer of trust.

---

[65] Install Geth - https://geth.ethereum.org/install-and-build/Installing-Geteth, accessed on 2019-05-28
[66] Private Network - https://github.com/ethereum/go-ethereum/wiki/Private-network, accessed on 2019-05-28
[67] Private Network - https://medium.com/@yashwanthvenati/setup-private-ethereum-blockchain-network-with-multiple-nodes-in-5-mins-708ab89b1966, accessed on 06-15-2019
[68] Ethereum - https://medium.com/cybermiles/running-a-quick-ethereum-private-network-for-experimentation-and-testing-6b1c23605bce, accessed on 08-03-2019
[69] Example Private Ethereum - https://medium.com/coinmonks/private-ethereum-by-example-b77063bb634f, accessed on 2019-08-24
[70] Example Infura - https://www.freecodecamp.org/news/hands-on-get-started-with-infura-and-ipfs-on-ethereum-b63635142af0/, accessed on 09-04-2019

This test confirms the viability of this solution for systems or applications that generate log files with alarming or monitoring events.

2.6.2. 2nd Iteration in the proposed methodology

The evaluation of the 1st iteration made it possible to verify that the records could be manipulated in the generation of the log file or in the insertion in the database.

The solution proposed in this 2nd iteration of the methodology allows improving the integrity of the data, maintaining the data flow already demonstrated in the first iteration by guaranteeing that the data at the time of registration in the database cannot be manipulated, through secure registrations, registration to registration, by the respective hash preventing internal attacks.

The security of data records, in the 2nd iteration, aims to guarantee at the time of event registration that the records are available at the accepted nodes of this private network of different entities without the possibility of tampering.

This artifact is focused on avoiding intrusion or manipulation attempts that could compromise the entry of the record and condition the generation of the log file or access to the log files (1st iteration) avoiding falsifying the complete record, truncating the log, inventing new records and inject past registry entries.

THE Figure III.9shows which object is planned for the 2nd iteration and which concerns the phase of inserting events into the database from IoT equipment, local and operator data concentrators. The event insertion, presented below, consolidates the different data flows taking into account the different endpoints.

Figure III.9 - Phase of the Insertion Data Flow in the Database

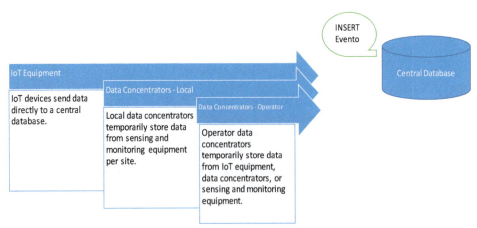

This phase uses untreated data from equipment and systems, which are integrated into one or more databases. The communication mechanisms can use the publication and subscription mechanisms with the exchange of messages through the MQTT protocol or with the sending of xml files with predefined structures and formats.

The proofs of concept developed did not focus on the phase of data flow from IoT devices to the database and on the way in which all events are subject to mechanisms for prior verification of the authenticity of the equipment, to which data fields would be added control.

The artifact begins at the stage of the insertion process in the database, of data originating from IoT equipment, which from there creates the blocks that contain the control data of sets of records, using the private blockchain network and the trees Merkel to group them by hash matching.

The joint validation of records thus makes it possible to create a control network over sets of event records, optimizing validation and consensus mechanisms. The Figure III.10 presents this flow and the implementation of a trigger that adds the hash key of the events to be inserted in the database.

Figure III.10 - Insertion with per-register hash generation

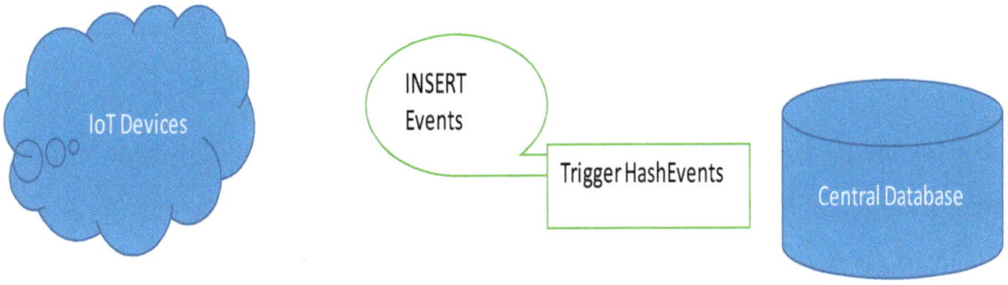

The insertion phase is still concentrated in a central point to facilitate the proof of concept. In this scenario, we chose to include a trigger, referred to in the previous figure, which creates the control data (hash of the record) that will serve as the basis for the blockchain network. The use of Merkel's tree allows validating groups of records, improving the performance of the solution. The blockchain network, in this scenario, constitutes a control network that validates the trust of the data. As in the 1st iteration of the methodology, the blockchain network serves mainly to distribute control data and access to event data, constituting a layer of trust.

This scenario, more detailed in point2.6.2.1 with the Operationalization of the 2nd Iteration, is presented in Figure III.11 where the limitation still remains, that the event data resides in central databases, which contain the data accessed through the control data of the event logs, on the blockchain network.

Figure III.11 - Artifact demonstration scenario (2nd iteration)

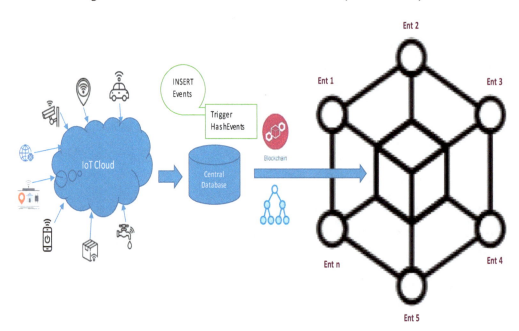

This iteration tested and validated the data flow control at the record level, for the type of blockchain application (BC3-Data flow control) foreseen in the model.

It was not possible to test and evaluate the other types of blockchain application (BC4-Acceptance of devices, BC5-Version control and BC6-Security of systems) foreseen in the model, to guarantee the authenticity of the equipment, the correct version installed and security of the system. The tested ticketing network equipment does not allow this implementation, as it does not have the functionality of managing private and public keys or certification to guarantee the integrity of the data and the storage is limited to store the data of the nodes of the associated blockchain network. to each device and in offline situations.

2.6.2.1. Operationalization of the 2nd Iteration

The 2nd Iteration involved accessing the data model and changing the data structure of the system or application, thus being intrusive on existing systems.

In this simulation, the following trigger was added, which in the phase of inserting records into the database adds a HashEvent control field that allows guaranteeing the integrity of the data of each record and is the value that is inserted in the block to constitute the control blockchain about event logs.

```
CREATE TRIGGER dbo.HashEvents
ON events
INSTEAD OF INSERT
AT
BEGIN
 SET NOCOUNT ON;

 INSERT dbo.Events (field 1, field 2, ..-, field n, HashEvent)
 SELECT field 1, field 2, ..., field n,
 HASHBYTES('SHA2_256', CONCAT([field 1],'|',[field 2],'|', …,'|',[field n]))
 FROM inserted;
END
```

Fields 1 to n are at least those referred to in Table III.9 - Fields exported to the Log file.

The simplified data structure of the blockchain is described in the following figure:

Figure III.12 - Simplified Data Structure of Blocks in the 2nd Iteration

The blockchain in this version has a block for each record, with fields that allow access to the EventID, AccessDB data, and the HashEvent (record control field) and EventDateHour (event date/time) control fields.

The amount of predictable records in a multi-operator ticketing system with thousands of devices sending events from the numerous signals will imply the search for solutions that allow the validation of large amounts of records/transactions.

The proposed mechanism is based on Merkel trees in which each hash of the record serves as input to the Merkle tree which, together with other records, recursively recalculates the MerkleRoot value.

The MerkleRoot value is inserted in a single block to allow massive data validation. The process starts with inserting the hash of each record into a separate block that is added to the next record, resulting in a single block, until a single block is output.

THE Figure III.13 presents the data structure with Merkel tree and the way the blocks are related.

Figure III.13 - Data structure with Merkel tree

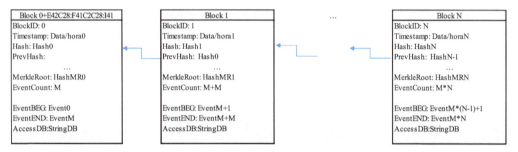

Merkle trees validate a set of transactions. The N data records are combined into a Merkle tree. To check if a certain element is included in the tree, a maximum of 2 * log2 (N) calculations are required. This algorithm provides an efficient way of checking whether a transaction is in a block or associated with a block.

The test blockchain network remained with three nodes and was developed by adding to the blockchain base data structure the hash resulting from the Merkle tree root, which also allows validating the sets of data records.

This proof of concept treats a block of events, through an algorithm to generate a hash, which checks the validity of the dataset based on the original events. The hash function is not executed all at once for the set of events, but results from the hash of each event being then gathered into a hash resulting from the implementation of the Merkle tree.

The structure looks like a tree, where the base nodes contain N events. Each hash pair is combined from the bottom row, through intermediate hashing to the top hash, which allows validating the entire set of records.

2.7. Assessment (5th Step DSRM methodology)

The evaluation results from the demonstration of the proofs of concept step by step and the verification that the logs obtained and transmitted prove to be immune to tampering or manipulation and that their delivery is guaranteed exclusively to authorized recipients (1st Iteration). The data from the file link in the IPFS network and the respective hash were inserted into the blockchain network, constituting a layer of control over the access data to the log files, guaranteeing the trust of the data and its integrity.

In the proof of concept, of the 2nd iteration, the event data transmitted by each device, in real time, offline or in blocks, made it possible to insert the event records, with mechanisms of authenticity and control over the fields of the data records. .

The control data allowed access to the data and ensured the integrity of the data contained in the records validated by hash functions and which were inserted into the blockchain network to constitute the layer of control over the recorded data.

This evaluation process began by reviewing and comparing systems designed without this security concern and those that arise from their design supported by mechanisms based on blockchain technology.

The artifacts developed separated event data from control data to control the flow of data, ensuring that the logs sent to the IPFS network or the records contained in the databases are validated by the control data residing in the blockchain network.

This evaluation of the monitoring and alarming component of the ticketing system, in two iterations, made it possible to verify that the predicted proof-of-concept artifacts can guarantee data security and data flow control.

The iterative process foreseen in the methodology allowed to improve the data flow control and can be extensively adapted to improve the way we can guarantee the confidence in the good functioning of the system, creating a data flow control layer supported on blockchain technology, a layer reliable.

The most relevant limitations are having evaluated the alarmistic components and not the operations or transactions. A part of the data flow between the 5 levels defined in the model was also evaluated.

In a broader context, the model can see its application extended to other smart places, architectures and applications, as a model that can be applied to the development of similar systems.

2.8. Communication (6th DSRM Step)

The communication process was based on the writing of manuscripts on three essential aspects for the elaboration of this work embodied in this thesis, the systematic literature review, the generic reliable data model of a smart city and the data markets. For this purpose and in the course of the research and investigation work, we prepared three manuscripts that were presented at the 2018 and 2019 World Conference on Information Systems and Technologies (WordCIST) and 2018 International Conference on Software Process Improvement (CIMPS) conferences, respectively with the titles "Systematic Review of Literature, Research on Blockchain Technology as Support for the Proposed Trust Model Applied to Smart Places", published March 28, 2018, "Marketplace of Trusted Data",(Brandão et al., 2018b, 2018a, 2018a). These manuscripts are referenced in the specific section of Publications at the end of the thesis.

The contribution of this research work reveals the feasibility of the proposed generic data model, supported by blockchain technology and applied to a smart city. The proofs of concept developed were oriented towards validating the model, using blockchain technology, mainly in the control of data flow, data authenticity (logs and records) and data integrity, through a control layer, the trust layer.

2.9. Practical Results

Files can only be accessed by name or key and its path. The choice of IPFS (Inter-Planetary File System) allows data redundancy, transparency (security and authenticity) and decentralization that facilitates the storage of large amounts of data, avoiding the limitation of the exclusive use of the blockchain, in which the amount data is still a critical aspect.

The recording of system or application logs provides, asynchronously, the possibility of analyzing and evaluating the events that occurred and that were the object of a previous configuration of the audit items to be recorded.

The consolidated use of log files (from various systems or equipment) allows tools such as security event management and correlation systems (SIEM- Security Information and Event Management), with robust resources and machine learning for incident detection, its correlation and alert. It also allows you to obtain predictive analysis, notify, diagnose, send alarms and detect anomalies, for the answer to the problem, avoiding affecting the systems in operation or reducing the downtime of critical equipment.

The authenticity and integrity of the data is a critical aspect that can condition the responses and automatisms that can be triggered, many of them without human intervention. In this way, guaranteeing the reliability of the data is very critical to avoid manipulation, undue access to the generated logs, which serve as the basis for various analyzes and information for decision-making, which is often automated.

In addition to the choice of the IPFS distributed file system (supported on blockchain), blockchain technology was used to allow controlling the data flow of log files in a distributed, cryptographically protected way and verifying the correctness of the data, the verification of the log through the block on the blockchain that proves the accuracy of the logs.

For cryptographic security, we use asymmetric cryptography through the GPG (GNU Privacy Guard) tool, which allows implementing the key mechanisms (private and public), described in point2.6.1.1.

The blockchain, in the 1st iteration, serves as a pointer to the IPFS and to verify its hash. The hash generated when adding the file to the IPFS network can serve as a hash that guarantees the integrity of the generated log, however, the more robust hash (SHA512) can be added that will validate the respective block and allow:

- Obtain log access on the IPFS network;
- And the decryption of confidential files stored in IPFS protected by private and public keys using GPG.

The cryptographic hash function has the following characteristics: it cannot generate the same hash value for different inputs; the same data generates the same hash value; quickly generates a hash for any dataset; unable to calculate input based on hash value; and any change to the input data changes the hash.

The blockchain in this context serves as a control layer to ensure trust in the data contained in the logs available on the IPFS network and its decryption by safe use between nodes of the blockchain network.

In this context, the traffic management system logs (TT_Logs and Event_Logs) were also simulated, which are automatically generated by the proprietary system (TMS) and resulted in the same way as data flow control.

This approach proved to be less intrusive in existing systems and made it possible to guarantee the authenticity of the logs on any node on the network and the integrity of the data contained in the logs.

The 2nd iteration added a control layer implemented through the blockchain network to validate the records inserted in the databases and with them guarantee data integrity as soon as the events arrive, by the various equipment, at the record level.

This artifact leads to a higher level of confidence which makes the inserted records inviolable and with the use of Merkel trees it optimizes the validation of blocks of records improving the performance of the record validation system.

The integration of big data with blockchain technology may also allow us to answer and manage most of the main questions and challenges that involve big data, especially in terms of trust in the data that support it.

Blockchain technology is expected to give greater confidence in data integrity, immutable inputs, timestamps, consensus-based acceptance, audit trails, and assurance about the origin of data. These are the areas that could make blockchain technology central. In addition to data integrity, the shared data layer that blockchain technology can introduce creates an entirely new set of possibilities, capabilities and knowledge for Artificial Intelligence to act.(Rabah, 2018)

3. Discussion of Results

Analyzing the results of the tests carried out in the simulation of the two iterations, it is possible to discuss the potential of the developed solution.

This analysis demonstrates that this type of approach makes it possible to detect tampering with the logs, guarantee the authenticity of the logs and guarantee that the records inserted in the databases are not subject to manipulation, reducing failures by omission or manipulation of the data of the records and generated logs.

The 1st iteration verified through the implemented proofs of concept that it is possible to create a layer of trust through the blockchain network, allowing the data contained in the logs to be available and accessible in the IPFS network at any node of the trust network.

These proofs of concept also demonstrate that a developed system that uses log files as a basis for analysis or evaluations, using various applications or different entities, produces reliable data, supported by authentic logs and which are available to detect the events that constituted the basis for calculating unavailability, errors or problems that allow predictive analyzes or deviations from standards or confidence intervals for the proper execution of the processes and the main groups of functions described in Figure III.14, in an IoT environment.

Figure III.14 - Main Data Groups/Functions

The Figure III.14 thus reveals the main functions in an IoT environment that can be used by different datasets and that need to trust the data, through a layer of trust supported by blockchain and that are provided to present the correct information and that allows an objective analysis, proper management and correct functioning of the market and customer satisfaction.

From this 1st iteration we can say that this control process based on blockchain and IPFS can be used in systems or applications that generate logs that will be used by other applications and/or by other entities. The 2nd iteration adds this control to the record level, increasing data security.

In addition to the 2nd iterations carried out within the scope of this research work, we can see that this model could be extended to the previous stages, from IoT equipment directly to the blockchain network, having resolved the limitations of the sensing devices and IoT equipment still in operation, verified at the same time. resource level, namely processing, transmission, storage and functionalities of IoT equipment.

It should be noted that the proposed model was tested in the event components and not in the transactional components that involve monetization.

The proposed approach establishes the separation of the data layer from the control layer, which is effective through blockchain technology, adopting the necessary fields to define data access paths and data integrity checks.

The fields are adopted as the data flow to be controlled is analyzed (see fields adopted in the 1st iteration and in the 2nd iteration, at point 2.6.1.1 and at the point 2.6.2.1).

This control layer, which we call the trust layer, provides the nodes of the private blockchain network with the data for accessing and verifying the integrity of the data.

The model proposed at point 2.2 (Generic Data Model) of Chapter II bases its design on data exchanges mainly through data interoperability between ecosystems. Data management and data flow control need a layer of trust to allow the data to be unique and reliable, providing the smart city with harmonious dynamics between its ecosystems, the automation of processes and the possibility of establishing algorithms. of artificial intelligence and big data analysis to promote, in an incremental and well-tested way, new forms of decision making.

The characteristics of the control data must be analyzed and structured to allow the data flow to be consistent and sufficiently granular at each level to be able to segment the shared data with the appropriate protections and permissions.

The Figure III.15 presents an overview of the solution in which the cloud-based IoT network and IoT platforms would be controlled by a blockchain network and with it establish a control or trust network to manage the flow of data and ensure data trust through implementation of the types of blockchain applications (from Table III.7) defined at point 2.2, from ChapterIII, which go through the 6 defined types: Secure Transactions, Data Security, Data Flow Control, Device Acceptance, Version Control and Systems Security.

Figure III.15 - Data Flow considering an IoT network and a Blockchain network

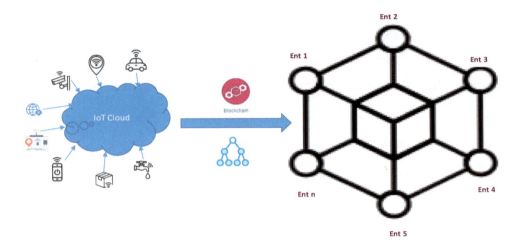

The Figure III.16 presents the objective scenario in which the blockchain network integrates with the IoT network and in this scenario the IoT control and data layers come together and are supported in the blockchain network. This scenario may have simplification or grouping mechanisms that facilitate processing, consensus, storage and transmission, as for example the proposed Chen, Wang, and Wang (2018).

Figure III.16 - Data Flow Integrating IoT with Blockchain

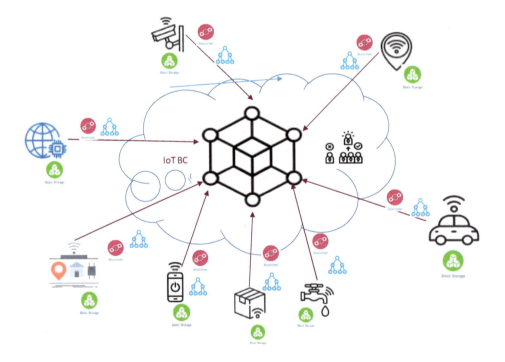

This scenario, as we have seen, still has several limitations with the difficulty of obtaining the storage and computation to handle large amounts of data on a blockchain network. Having constituted, in this investigation, the main reason for adopting blockchain technology as a trust or control layer, which may in the future include the data layer.

3.1. Related works

For command and control systems using distributed computing protocols, a proof of concept was developed using IPFS and IPNS that allows for more secure and anonymous communications. (From Aquino et al., 2018)

Cucurull & Puiggalí (2016) presented an implementation solution using blockchain technology with the underlying bitcoin framework to increase the security of immutable logs. The PoW consensus mechanism that supports the blockchain project, based on the structure used in the bitcoin network, which cannot be modified by the intervention of less than 50% of the mining capacity of the validating nodes.

Ali, Dolui, & Antonelli (2017) present a blockchain software stack and the peer-to-peer interplanetary file system (IPFS) to control decentralized access and the blockchain to create an operations leader of IoT data, with several limitations namely the difficulty in maintaining a flow of data, data through IPFS and exploit without blockchain to increase block processing time and storage savings on validator nodes.

The design and implementation presented by Hasan, Sultan, & Barbhuiya (2019) provides a decentralized solution without cloud storage provider (CSSP-Cloud Storage Provider) price control to provide blockchain and IPFS based cloud storage service for data validation.

IPFS architecture has been improved (Y. Chen et al., 2017) to allow more income for individual users and content providers, so that the IPFS information of each node can be saved on the blockchain and for this, it combines three forms of replication and storage in this scheme.

Zheng, Li, Chen, & Dong (2018) design an IPFS-based storage model for blockchain for various transaction types, looking to improve storage space, security, and node synchronization.

The review of a global mechanism to protect IoT layers, taking into account the diversity of IoT resources, was presented by Khan & Salah (2018), where IoT security was analyzed and the issues were categorized according to the IoT layers (high-level, intermediate-level, and low-level). In this work, they analyzed the use of blockchain technology for some of the main IoT security problems. The application of blockchain technology, supported by smart contracts, will allow the management, control and protection of IoT devices, using the intrinsic resources of the blockchain for IoT security, namely, address space, identity of things and their governance, authentication and data integrity, authentication, authorization and privacy, and secure communications.

In summary, it appears that there are several works that rely on the IPFS network and use the blockchain network to solve some of the problems encountered in terms of security, authenticity and data integrity, strengthening the contribution of this work in the deepening of knowledge and in the how smart places can prove to be more reliable and with reliable data.

IV. FINAL CONSIDERATIONS

IV. FINAL CONSIDERATIONS

Smart places and, in this specific case, smart cities, support their operation in a vast set of data that monitor and act on the various ecosystems.

The smart city is physical, digital and virtual, encompassing several factors that include citizen participation, the participation of economic and social agents, governance, innovative policies, technological infrastructure, ecological, energy, social and economic sustainability, and mobility and transport policies, which converge on dynamic strategies and new approaches.

Participation and interaction strategies require robust and resilient connectivity infrastructures, which support the multiplicity of equipment, sensors and actuators, which generate huge amounts of data, which are processed automatically to extract relevant information, to reorient strategies, policies and the creation of ecosystems that foster new business models.

In this context, trust in data plays a fundamental role in the security, dynamics and functioning of the smart city.

Transport networks, as critical infrastructures in the flows of people and goods, are supported by different systems with different degrees of maturity and innovation. The systems are mainly legacy systems with proprietary solutions. The solutions found have to take into account this scenario and the need to comply with lengthy testing and certification processes.

The scientific methodology, DSRM(Peffers et al., 2007), adopted in this research work, proved to be adequate and allowed the dynamics that served as a basis for the exploration of artifacts that could demonstrate and validate the proposed generic data model.

The option was for a system transversal to the mobility and transport ecosystem such as the ticketing system, in the alarm and monitoring component, and that had no restrictions on access to its data model and data structure, important for the 2nd iteration. The approach followed in this work proved to be less intrusive in systems and applications, adding a layer of trust.

The proposed model aimed to increase the protection of infrastructures, systems and applications against data tampering and omission of events, reducing the possibility of undue access that alters the data entered.

This work allowed, in the scope of smart cities, in IoT environments, to propose a generic data model supported by six types of blockchain application and that, through the artifacts based on proofs of concept, demonstrated the feasibility of guaranteeing the authenticity of log files and the its integrity, and the integrity of the event logs.

Both iterations revealed the possibility of creating a layer of trust that allows ensuring data integrity and authenticity of data origin, based on the use of blockchain technology and the IPFS network supported by blockchain.

In the introduction, the general objective and three specific objectives were identified. The research work made it possible to achieve the objectives, which we seek to present in the following paragraphs and respond with the contributions achieved.

The general objective of the project is to apply a generic data model, to be proposed, in support of the smart city concept, in order to systematize its actions and the control of data flows and data quality, which allow the management of data. and information, reliably and securely.

In conclusion, it was possible to propose a generic model of data to support the smart city (point 2.2, of chapter III) that systematizes and allows the management of data and information, in a reliable and secure way. The evaluation of its viability results from the validation of the artifacts confirming the control of the data flow in the treatment of logs and in the insertion of new data records.

In summary and below, the following three specific objectives are defined that were intended to be achieved with the completion of the work to be developed.

1. To structure a generic data model to support the smart city concept that leads and allows the alignment of the application of data ecosystems with natural ecosystems.

This objective was achieved in the design of the generic data model, with the organization of the smart city in ecosystems that relate to and are "powered" by IoT data, grouped by application domains and supported by 6 types of blockchain applications, to ensure the authenticity and integrity of the data.

2. Structuring the relationships between ecosystems, participants and data that facilitate the use of blockchain technology in data management.

The creation of a layer of trust adapted to different applications allows data management, regardless of the data structure and data location.

3. Ensure mechanisms of reliability in the management of data and data sources.

The proposed reliability mechanisms are revealed in the six types of application of blockchain technology that support the functioning of the smart city and its ecosystems, providing the set of functionalities that allow managing and controlling data, the authenticity of data sources and the origin of IoT equipment.

1. Conclusions

Smart places reveal physical spaces, of varying dimensions and with Organizations that manage these spaces, which are dematerialized into digital and virtual spaces, enhancing citizen participation, optimizing the management of resources, improving the quality of spaces and citizens, and providing levers for innovation, social inclusion, environmental, energy and economic sustainability.

The massive use of IoT and mobile equipment, integrated in the Cloud and accessed via the Internet, supported by robust connectivity infrastructures, enhances real-time monitoring, dynamic evaluation, response and actuation algorithms, machine-to-machine, artificial intelligence, data mining algorithms, big data processing, machine learning, etc.

This reality, which is becoming more dense leads to new problems, with the amount of data generated and the difficulty in controlling data flows, data quality and confidence in the origin of the data.

The problem of controlling data flows, data quality and trust in the origin of data, in smart cities, focuses the research work that was developed.

In conclusion, we can verify that the artifacts based on proofs of concept allowed us to confirm that blockchain technology can constitute a layer of trust over the data and in this way guarantee the authenticity of the data and the integrity of the data.

The 1st iteration allowed controlling the flow of data files, log files, guaranteeing the authenticity of the files and the integrity of the data contained in the files and also allowed controlled access through the encryption and decryption mechanism, through private and public keys.

The central object of the work focused on the alarming and monitoring components of the systems, resulting in logs that extract and store events, where problems, alarms, incidents, availability or unavailability of equipment, services and components, security breaches, access attempts, improper access and many other events that may imply predictive, preventive or corrective actions, behaviors and analyzes and consequent operation and maintenance actions.

The result of the work showed that this model is viable and the solutions are universal for systems and applications that manage log files, proving in the control of the data flow of alarming and monitoring of the ticketing system and the traffic management system. These artifacts confirmed the application of the proposed model in the type of blockchain application mainly in data flow control (BC3-Model data flow control), not being intrusive in systems and applications.

The 2nd iteration added to the first process of controlling the data flow of the log files, the control of the event records inserted in the database. Blockchain technology has also made it possible to constitute a layer of trust that guarantees data authenticity and data integrity, and in this way provides the mechanism for accessing data in a controlled manner and with the guarantee of non-tampering of data. This iteration could be improved with the use of Merkle trees that allowed validating groups of records with the hash resulting from the combination of the hash of the base records, up to the root, optimizing the processing and storage in view of the treatment from record to record.

This iteration necessarily implied access to the data model of the applications, being more intrusive, implying the addition of control and data access fields, and activating, in the insertion phase, triggers that provided the mechanisms of control and construction of the blockchain of records. As in the 1st iteration, blockchain technology was constituted as a layer of data trust, maintaining control information through hashs and data access information (links).

At this stage and with the research work carried out, we can say that blockchain technology guarantees data authenticity and data integrity, through a layer of trust, which links it to the data, but is independent of the data layer.

The study carried out allows us to envisage the growing evolution in terms of research on blockchain technology, supported by the huge investment that is made in several areas and on various aspects of blockchain technology, which leads to the expectation that new business models will tend to use "smart" contracts. " dynamic and decentralized structures, with Value propositions based on blockchain technology and will have to fulfill the following basic characteristics: immutability, encryption, distribution, tokenization and decentralization.

Blockchain technology will tend to incorporate or be incorporated into complementary technologies such as decentralized identity (SSI-Self-Sovereign Identity), big data, IoT and artificial intelligence (AI), providing the layer of trust and, in the future, with the Expected processing improvements, with more efficient consensus, and with greater storage capacities, will include the data layer.

2. Limitations and Future Work

This work did not study the monetized transactional systems that support various activities inherent to the functioning of a smart city.

The proposed model was evaluated mainly in data flow control and resulted in the adoption of solutions that separate the data layer from the control layer, which was called the trust layer. This limitation results from the difficulty in obtaining storage capacity and distributed computing to handle large amounts of data, as is the case, through a blockchain network that includes large volumes of IoT data.

Future work should focus on transactional systems and complete the remaining phases of the data flow from IoT equipment, to data storage and processing, with the creation of layers of trust, supported on a blockchain network that provides data security.

Work should also be carried out for the remaining types of blockchain applications defined in the model, namely verifying that the IoT equipment is valid equipment and performing version control by type of equipment, and from there, being able to verify the security of the transmitted data, in real-time or in blocks, allowing the existence of blocks of offline events in some phases of the sending.

Bibliography

Abella, A., Ortiz-de-Urbina-Criado, M., & De-Pablos-Heredero, C. (2017). A model for the analysis of data-driven innovation and value generation in smart cities' ecosystems. Cities, 64, 47–53. https://doi.org/10.1016/j.cities.2017.01.011

Agarwal, A., Dahleh, M., & Sarkar, T. (2018). A Marketplace for Data: An Algorithmic Solution. ArXiv:1805.08125 [Cs]. http://arxiv.org/abs/1805.08125

Ai B, Guan K, Rupp M, Kurner T, Cheng X, Yin X-F, Wang Q, Ma G-Y, Li Y. , Xiong, L., & Ding, J.-W. (2015). Future railway services-oriented mobile communications network. IEEE Communications Magazine, 53(10), 78–85. https://doi.org/10.1109/MCOM.2015.7295467

Aken, JE van. (2004). Management Research Based on the Paradigm of the Design Sciences: The Quest for Field-Tested and Grounded Technological Rules: Paradigm of the Design Sciences. Journal of Management Studies, 41(2), 219–246. https://doi.org/10.1111/j.1467-6486.2004.00430.x

Al Nuaimi, E., Al Neyadi, H., Mohamed, N., & Al-Jaroodi, J. (2015). Applications of big data to smart cities. Journal of Internet Services and Applications, 6(1). https://doi.org/10.1186/s13174-015-0041-5

Alahadhi, S., & Scholl, HJ (2016). Smart Governance: A Cross-Case Analysis of Smart City Initiatives. 2016 49th Hawaii International Conference on System Sciences (HICSS), 2953–2963. https://doi.org/10.1109/HICSS.2016.370

Albino, V., Berardi, U., & Dangelico, RM (2015). Smart Cities: Definitions, Dimensions, Performance, and Initiatives. Journal of Urban Technology, 22(1), 3–21. https://doi.org/10.1080/10630732.2014.942092

Ali, MS, Dolui, K., & Antonelli, F. (2017). IoT data privacy via blockchains and IPFS. Proceedings of the Seventh International Conference on the Internet of Things - IoT '17, 1–7. https://doi.org/10.1145/3131542.3131563

Alter, M. (2015). Best Practices for IT Service Management. Introducing ITIL. https://www.nysforum.org/events/4_30_2015/Final.pdf

Alturki, A., Gable, GG, & Bandara, W. (2011). The Design Science Research Roadmap. In H. Jain, AP Sinha, & P. Vitharana (Eds.), Service-Oriented Perspectives in Design Science Research (Vol. 6629, pp. 107–123). Springer Berlin Heidelberg. https://doi.org/10.1007/978-3-642-20633-7_8

Amarnath, A. (2011). City as a Customer Strategy: Growth Opportunities From The Cities of Tomorrow. 31.

Amjad, S., & Javaid, N. (2019). The Secure Service System for Clients through Service Provider by using the IoT in Blockchain. 7.

Amorim, AL de. (2016). Smart Cities and City Information Modeling. XX Congress of the Iberoamerican Society of Digital Graphics, 481–488. https://doi.org/10.5151/despro-sigradi2016-440

Anjum, A., Sporny, M., & Sill, A. (2017). Blockchain Standards for Compliance and Trust. IEEE Cloud Computing, 4(4), 84–90.

Anthopoulos, LG (2015). Understanding the Smart City Domain: A Literature Review. In A. de Janvry & R. Kanbur (Eds.), Poverty, Inequality and Development (Vol. 1, pp. 9–21). Springer US. https://doi.org/10.1007/978-3-319-03167-5_2

Anthopoulos, LG, Janssen, M., & Weerakkody, V. (2015). Comparing Smart Cities with different modeling approaches. Proceedings of the 24th International Conference on the World Wide Web, 525–528. http://dl.acm.org/citation.cfm?id=2743920

Anwar, U. (2017). Blockchain: Anonymisation Techniques within Distributed Ledgers. 6.

Arasteh, H., Hosseinnezhad, V., Loia, V., Tommasetti, A., Troisi, O., Shafie-khah, M., & Siano, P. (2016). Iot-based smart cities: A survey. Environment and Electrical Engineering (EEEIC), 2016 IEEE 16th International Conference on, 1–6.

Adverton, D., & Fitzgerald, G. (2006). Information systems development: Methodologies, techniques and tools (4th ed.). McGraw Hill.

Babich, V., & Hilary, G. (2019). Distributed Ledgers and Operations: What Operations Management Researchers Should Know about Blockchain Technology. 39.

Bach, LM, Mihaljevic, B., & Zagar, M. (2018). Comparative analysis of blockchain consensus algorithms. 2018 41st International Convention on Information and Communication Technology, Electronics and Microelectronics (MIPRO), 1545–1550. https://doi.org/10.23919/MIPRO.2018.8400278

Badii, C., Bellini, P., Cenni, D., Difino, A., Nesi, P., & Paolucci, M. (2017). Analysis and assessment of a knowledge based smart city architecture providing service APIs. Future Generation Computer Systems, 75, 14–29. https://doi.org/10.1016/j.future.2017.05.001

Baliga, DA (2017). Understanding Blockchain Consensus Models. 14.

Baskerville, R., Pries-Heje, J., & Venable, J. (2009). Soft design science methodology. Proceedings of the 4th International Conference on Design Science Research in Information Systems and Technology - DESRIST '09, 1. https://doi.org/10.1145/1555619.1555631

Bates, O., & Friday, A. (2017). Beyond Data in the Smart City: Repurposing Existing Campus IoT. IEEE Pervasive Computing, 16(2), 54–60.

Bauer, I., Zavolokina, L., Leisibach, F., & Schwabe, G. (2019). Exploring Blockchain Value Creation: The Case of the Car Ecosystem. 10.

Bellini, P., Nesi, P., & Pantaleo, G. (2015). Benchmarking RDF Stores for Smart City Services. 2015 IEEE International Conference on Smart City/SocialCom/SustainCom (SmartCity), 46–49. https://doi.org/10.1109/SmartCity.2015.45

Benet, J. (2014). IPFS - Content Addressed, Versioned, P2P File System. ArXiv:1407.3561 [Cs]. http://arxiv.org/abs/1407.3561

Benevolo, C., Dameri, RP, & D'Auria, B. (2016). Smart Mobility in Smart City. In T. Torre, AM Braccini, & R. Spinelli (Eds.), Empowering Organizations (Vol. 11, pp. 13–28). Springer International Publishing. https://doi.org/10.1007/978-3-319-23784-8_2

Beyer, C., Elisei, P., Popovich, VV, Schrenk, M., & Zeile, P. (2015). REAL CORP 2015. Plan Together—Right Now—Overall. From Vision to Reality for Vibrant Cities and Regions Proceedings of the 20th International Conference on Urban Planning, Regional Development and Information Society.

Bharadwaj, AS, Rego, R., & Chowdhury, A. (2016). IoT based solid waste management system: A conceptual approach with an architectural solution as a smart city application. 2016 IEEE Annual India Conference (INDICON), 1–6. https://doi.org/10.1109/INDICON.2016.7839147

Biswas, K., & Muthukkumarasamy, V. (2016). Securing Smart Cities Using Blockchain Technology. 2016 IEEE 18th International Conference on High Performance Computing and Communications; IEEE 14th International Conference on Smart City; IEEE 2nd International Conference on Data Science and Systems.

Bo Chen, & Cheng, HH (2010). A Review of the Applications of Agent Technology in Traffic and Transportation Systems. IEEE Transactions on Intelligent Transportation Systems, 11(2), 485–497. https://doi.org/10.1109/TITS.2010.2048313

Bongaerts, R., Kwiatkowski, M., & König, T. (2017). Disruption Technology in Mobility: Customer Acceptance and Examples. In A. Khare, B. Stewart, & R. Schatz (Eds.), Phantom Ex Machina (pp. 119–135). Springer International Publishing. https://doi.org/10.1007/978-3-319-44468-0_8

Boyes, H. (2016). Cyber security risks in the Built Environment—Standards, Skills & Apprenticeships.

Braem, B., Latre, S., Leroux, P., Demeester, P., Coenen, T., & Ballon, P. (2016). Designing a smart city playground: Real-time air quality measurements and visualization in the City of Things testbed. Smart Cities Conference (ISC2), 2016 IEEE International, 1–2.

Brandão, A., Mamede, HS, & Gonçalves, R. (2018a). Smart City's Model Secured by Blockchain. In J. Mejia, M. Muñoz, Á. Rocha, A. Peña, & M. Pérez-Cisneros (Eds.), Trends and Applications in Software Engineering (Vol. 865, pp. 249–260). Springer International Publishing. https://doi.org/10.1007/978-3-030-01171-0_23

Brandão, A., Mamede, HS, & Gonçalves, R. (2018b). Systematic Review of the Literature, Research on Blockchain Technology as Support to the Trust Model Proposed Applied to Smart Places. in A. Rocha, H. Adeli, LP Reis, & S. Costanzo (Eds.), Trends and Advances in Information Systems and Technologies (Vol. 745, pp. 1163–1174). Springer International Publishing. https://doi.org/10.1007/978-3-319-77703-0_113

Brandão, A., Mamede, HS, & Gonçalves, R. (2019). Trusted Data's Marketplace. in A. Rocha, H. Adeli, LP Reis, & S. Costanzo (Eds.), New Knowledge in Information Systems and Technologies (Vol. 930, pp. 515–527). Springer International Publishing. https://doi.org/10.1007/978-3-030-16181-1_49

Brown, BC (2005). Theory and Practice of Integral Sustainable Development. 1(2), 39.

Bumanis, N., Vitols, G., Arhipova, I., & Mozga, I. (2017, May 24). Mobile ticket lifecycle management: Case study of public transport in Latvia. 16th

International Scientific Conference Engineering for Rural Development. https://doi.org/10.22616/ERDev2017.16.N015

Cachin, C. (2016). Architecture of the Hyperledger Blockchain Fabric. 4.

Cachin, C., Schubert, S., & Vukolić, M. (2016). Non-determinism in Byzantine Fault-Tolerant Replication. ArXiv:1603.07351 [Cs]. http://arxiv.org/abs/1603.07351

Cao, T.-D., Pham, T.-V., Vu, Q.-H., Truong, H.-L., Le, D.-H., & Dustdar, S. (2016). MARSA: A Marketplace for Realtime Human Sensing Data. ACM Transactions on Internet Technology, 16(3), 1–21. https://doi.org/10.1145/2883611

Casino, F., Dasaklis, TK, & Patsakis, C. (2018). A systematic literature review of blockchain-based applications: Current status, classification and open issues. Telematics and Informatics. https://doi.org/10.1016/j.tele.2018.11.006

Castro, M., & Liskov, B. (2002). Practical byzantine fault tolerance and proactive recovery. ACM Transactions on Computer Systems, 20(4), 398–461. https://doi.org/10.1145/571637.571640

Castro, M., & Liskov, B. (1999). Practical Byzantine fault tolerance. OSDI, 99, 173–186.

Ceballos, GR, & Larios, VM (2016). A model to promote citizen driven government in a smart city: Use case at GDL smart city. Smart Cities Conference (ISC2), 2016 IEEE International, 1–6. http://ieeexplore.ieee.org/abstract/document/7580873/

Cekerevac, Z., Prigoda, L., & Maletic, J. (2018). Blockchain Technology and Industrial Internet of Things in the Supply Chains. 6(2), 9.

Chakrabarti, A., & Chaudhuri, AK (2017). Blockchain and its Scope in Retail. 04(07), 4.

Chakrabarty, S., & Engels, DW (2016). A secure IoT architecture for Smart Cities. 2016 13th IEEE Annual Consumer Communications & Networking Conference (CCNC), 812–813. https://doi.org/10.1109/CCNC.2016.7444889

Chen, T., Wang, H., Ning, B., Zhang, Y., Tang, T., & Li, K. (2018). Architecture Design of a Novel Train-centric CBTC System. 2018 International Conference on Intelligent Rail Transportation (ICIRT), 1–5. https://doi.org/10.1109/ICIRT.2018.8641603

Chen, Y., Li, H., Li, K., & Zhang, J. (2017). An improved P2P file system scheme based on IPFS and Blockchain. 2017 IEEE International Conference on Big Data (Big Data), 2652–2657. https://doi.org/10.1109/BigData.2017.8258226

Chen, Y.-J., Wang, L.-C., & Wang, S. (2018). Stochastic Blockchain for IoT Data Integrity. 14.

Chepurnoy, A., Larangeira, M., & Ojiganov, A. (2016). Rollerchain, a Blockchain With Safely Pruneable Full Blocks. ArXiv:1603.07926 [Cs]. http://arxiv.org/abs/1603.07926

Chilipirea, C., Petre, A.-C., Groza, L.-M., Dobre, C., & Pop, F. (2017). An integrated architecture for future studies in data processing for smart cities. Microprocessors and Microsystems, 52, 335–342. https://doi.org/10.1016/j.micpro.2017.03.004

Chin, J., Callaghan, V., & Lam, I. (2017). Understanding and personalizing smart city services using machine learning, The Internet-of-Things and Big Data. 2017 IEEE 26th International Symposium on Industrial Electronics (ISIE), 2050–2055. https://doi.org/10.1109/ISIE.2017.8001570

Cho, H. (2018). ASIC-Resistance of Multi-Hash Proof-of-Work Mechanisms for Blockchain Consensus Protocols. IEEE Access, 6, 66210–66222. https://doi.org/10.1109/ACCESS.2018.2878895

Cho, S., Park, SY, & Lee, SR (2017). Blockchain Consensus Rule Based Dynamic Blind Voting for Non-Dependency Transaction. International Journal of Grid and Distributed Computing, 10(12), 93–106. https://doi.org/10.14257/ijgdc.2017.10.12.09

Chowdhary, N., & Deep Kaur, P. (2016). Addressing the characteristics of mobility models in IoV for smart city. 2016 International Conference on Computing, Communication and Automation (ICCCA), 1298–1303. https://doi.org/10.1109/CCAA.2016.7813919

Coccoli, M., Maresca, P., Stanganelli, L., & Guercio, A. (2015). An experience of collaboration using a PaaS for the smarter university model. Journal of Visual Languages & Computing, 31, 275–282. https://doi.org/10.1016/j.jvlc.2015.10.014

Cohen, B. (2013). Smart city wheel. Retrieved from SMART & SAFE CITY: http://www. smartcircle. org/smartcity/blog/boyd-cohen-the-smart-city-wheel.

Colding, J., & Barthel, S. (2017). An urban ecology critique on the "Smart City" model. Journal of Cleaner Production, 164, 95–101. https://doi.org/10.1016/j.jclepro.2017.06.191

Corak, BH, Okay, FY, Guzel, M., Murt, S., & Ozdemir, S. (2018). Comparative Analysis of IoT Communication Protocols. 2018 International Symposium on Networks, Computers and Communications (ISNCC), 1–6. https://doi.org/10.1109/ISNCC.2018.8530963

Corman, F., & Meng, L. (2015). A Review of Online Dynamic Models and Algorithms for Railway Traffic Management. IEEE Transactions on Intelligent Transportation Systems, 16(3), 1274–1284. https://doi.org/10.1109/TITS.2014.2358392

Cruz, NF da, & Marques, RC (2014). Scorecards for sustainable local governments. Cities, 39, 165–170. https://doi.org/10.1016/j.cities.2014.01.001

Cucurull, J., & Puiggalí, J. (2016). Distributed Immutability of Secure Logs. In G. Barthe, E. Markatos, & P. Samarati (Eds.), Security and Trust Management (Vol. 9871, pp. 122–137). Springer International Publishing. https://doi.org/10.1007/978-3-319-46598-2_9

Cybersecurity in Smart Buildings(Collaborative Industry Perspective N. 9835–19; Cybersecurity in Smart Buildings). (2015). Frost & Sullivan. https://www.switchautomation.com/wp-content/uploads/2015/12/Cybersecurity-in-Smart-Buildings_-Discussion-Paper.pdf

Dameri, PR (2012). Searching for Smart City definition: A comprehensive proposal. INTERNATIONAL JOURNAL OF COMPUTERS & TECHNOLOGY, 11(5), 2544–2551. https://doi.org/10.24297/ijct.v11i5.1142

Dao, D., Alistarh, D., Musat, C., & Zhang, C. (2018). DataBright: Towards a Global Exchange for Decentralized Data Ownership and Trusted Computation. ArXiv:1802.04780 [Cs]. http://arxiv.org/abs/1802.04780

de Aquino, BMM, de Lima, MVL, de Oliveira, JPCM, & de Souza, CT (2018). IPFS and IPNS Protocols as a Means for Botnet Control: Proof of Concept. Proceedings of the Cybersecurity on Connected Devices Workshop (WSCDC-SBRC 2018), 1.

de Jong, M., Joss, S., Schraven, D., Zhan, C., & Weijnen, M. (2015). Sustainable–smart–resilient–low carbon–eco–knowledge cities; making sense of the multitude of concepts promoting sustainable urbanization. Journal of Cleaner Production, 109, 25–38. https://doi.org/10.1016/j.jclepro.2015.02.004

Deakin, M. (2011). The embedded intelligence of smart cities. Intelligent Buildings International, 3(3), 189–197. https://doi.org/10.1080/17508975.2011.579340

Deakin, M., & Al Waer, H. (2011). From intelligent to smart cities. Intelligent Buildings International, 3(3), 140–152. https://doi.org/10.1080/17508975.2011.586671

Decker, C., Seidel, J., & Wattenhofer, R. (2014). Bitcoin Meets Strong Consistency. ArXiv:1412.7935 [Cs]. http://arxiv.org/abs/1412.7935

Desouza, KC, & Flanery, TH (2013). Designing, planning, and managing resilient cities: A conceptual framework. Cities, 35, 89–99.

Dhungana, D., Engelbrecht, G., Parreira, JX, Schuster, A., Tobler, R., & Valerio, D. (2016). Data-driven ecosystems in smart cities: A living example from Seestadt Aspern. Internet of Things (WF-IoT), 2016 IEEE 3rd World Forum on, 82–87. http://ieeexplore.ieee.org/abstract/document/7845434/

Dorri, A., Kanhere, SS, Jurdak, R., & Gauravaram, P. (2017). Blockchain for IoT security and privacy: The case study of a smart home. 618–623. https://doi.org/10.1109/PERCOMW.2017.7917634

Draskovic, D., & Saleh, G. (2017). Decentralized data marketplace based on blockchain. 16.

Duarte, F., de Carvalho Figueiredo, F., Leite, L., & Alcides Rezende, D. (2014). A Conceptual Framework for Assessing Digital Cities and the Brazilian Index of Digital Cities: Analysis of Curitiba, the First-Ranked City. Journal of Urban Technology, 21(3), 37–48. https://doi.org/10.1080/10630732.2014.940709

Dustdar, S., Nastic, S., & Scekic, O. (2016). A Novel Vision of Cyber-Human Smart City. 2016 Fourth IEEE Workshop on Hot Topics in Web Systems and Technologies (HotWeb), 42–47. https://doi.org/10.1109/HotWeb.2016.16

Dworkin, MJ (2007). Recommendation for block cipher modes of operation: The CCM mode for authentication and confidentiality (NIST SP 800-38c; p. NIST SP 800-38c). National Institute of Standards and Technology. https://doi.org/10.6028/NIST.SP.800-38c

Elmaghraby, AS, & Losavio, MM (2014). Cyber security challenges in Smart Cities: Safety, security and privacy. Journal of Advanced Research, 5(4), 491–497. https://doi.org/10.1016/j.jare.2014.02.006

ENISA (2017). Information Sharing and Analysis Centers (ISACs). 51. https://doi.org/10.2824/549292

Evans, J. (2017). A Systems Approach to Predicting and Measuring Workload in Rail Traffic Management Systems. 16.

Fazio, M., Celesti, A., Puliafito, A., & Villari, M. (2015). Big Data Storage in the Cloud for Smart Environment Monitoring. Proceeded Computer Science, 52, 500–506. https://doi.org/10.1016/j.procs.2015.05.023

Ferreira, MC, Nóvoa, H., Dias, TG, & Cunha, JF e. (2014). A Proposal for a Public Transport Ticketing Solution based on Customers' Mobile Devices. Procedia - Social and Behavioral Sciences, 111, 232–241. https://doi.org/10.1016/j.sbspro.2014.01.056

Forbes, T. (1999). Prime clusters and Cunningham chains. Mathematics of Computation, 68(228), 1739–1748. https://doi.org/10.1090/S0025-5718-99-01117-5

Fourie, CJ, & Chimusoro, O. (2018). AN EXAMINATION OF THE RELATIONSHIP BETWEEN SUPPLY CHAIN MANAGEMENT PRACTICES AND BUSINESS PERFORMANCE: A CASE ANALYSIS OF A PASSENGER RAIL COMPANY. 12.

Gaetani, E., Aniello, L., Baldoni, R., Lombardi, F., Margheri, A., & Sassone, V. (2016). Blockchain-based Database to Ensure Data Integrity in Cloud Computing Environments. 10.

Gallego-Lopez, C., & Essex, J. (2016). Understanding risk and resilient infrastructure investment. Evidence on Demand. https://doi.org/10.12774/eod_tg.july2016.gallegolopezessex3

Garcia, AEG (2017). Competitive Intelligence and the Development of Dynamic Capabilities in Organizations. Ibero-American Strategy Journal, 16(01), 91–98. https://doi.org/10.5585/riae.v16i1.2439

Gaubatz, G., Kaps, J.-P., Ozturk, E., & Sunar, B. (2005). State of the Art in Ultra-Low Power Public Key Cryptography for Wireless Sensor Networks. Third IEEE International Conference on Pervasive Computing and Communications Workshops, 146–150. https://doi.org/10.1109/PERCOMW.2005.76

Ghannem, A., Hamdi, MS, Abdelmoez, W., & Ammar, HH (2015). A context model development process for smart city operations. Service Operations And Logistics, And Informatics (SOLI), 2015 IEEE International Conference on, 122–127. http://ieeexplore.ieee.org/abstract/document/7367605/

Gharaibeh, A., Salahuddin, MA, Hussini, SJ, Khreishah, A., Khalil, I., Guizani, M., & Al-Fuqaha, A. (2017). Smart Cities: A Survey on Data Management, Security, and Enabling Technologies. IEEE Communications Surveys & Tutorials, 19(4), 2456–2501. https://doi.org/10.1109/COMST.2017.2736886

Ghosh, H. (2018). Data Marketplace as a Platform for Sharing Scientific Data. In UM Munshi & N. Verma (Eds.), Data Science Landscape (Vol. 38, pp. 99–105). Springer Singapore. https://doi.org/10.1007/978-981-10-7515-5_7

Ghuli, P., Kumar, UP, & Shettar, R. (2017). A Review on Blockchain Application for Decentralized Decision of Ownership of IoT Devices. 8.

Giang, NK, Lea, R., Blackstock, M., & Leung, VCM (2016). On Building Smart City IoT Applications: A Coordination-based Perspective. Proceedings of the 2nd International Workshop on Smart - SmartCities '16, 1–6. https://doi.org/10.1145/3009912.3009919

Gilad-Bachrach, R., Laine, K., Lauter, K., Rindal, P., & Rosulek, M. (2017). Secure Data Exchange: A Marketplace in the Cloud. 30.

Glaeser, EL (2006). Why Are Smart Places Getting Smarter? 4.

Grave, KM (2016). A Case Study of Smart Cities: The Role of Stakeholder Commitment. 221.

Gregor, S., & Hevner, AR (2013). Positioning and Presenting Design Science Research for Maximum Impact. MIS Quarterly, 37(2), 337–355. https://doi.org/10.25300/MISQ/2013/37.2.01

Guo, H., Wei Sim, JZ, Veeravalli, B., & Lu, J. (2018). Protecting Train Balise Telegram Data Integrity. 2018 21st International Conference on Intelligent Transportation Systems (ITSC), 806–811. https://doi.org/10.1109/ITSC.2018.8569616

Gupta, SS (2017). BLOCKCHAIN - The foundation behind Bitcoin. John Wiley & Sons, Inc.

Gurnik, P. (2016). Next Generation Train Control (NGTC): More Effective Railways through the Convergence of Main-line and Urban Train Control Systems. Transportation Research Procedia, 14, 1855–1864. https://doi.org/10.1016/j.trpro.2016.05.152

Hammer, S. (2018). The Blockchain Ecosystem. 39.

Hammi, MT, Hammi, B., Bellot, P., & Serhrouchni, A. (2018). Bubbles of Trust: A decentralized blockchain-based authentication system for IoT. Computers & Security, 78, 126–142. https://doi.org/10.1016/j.cose.2018.06.004

Hankerson, DR, Vanstone, SA, & Menezes, AJ (2003). Guide to elliptic curve cryptography. Springer.

Harrison, C., & Donnelly, IA (2011). The theory of smart cities. Proceedings of the 55th Annual Meeting of the ISSS-2011, Hull, UK, 55. http://journals.isss.org/index.php/proceedings55th/article/view/1703

Hasan, SS, Sultan, NH, & Barbhuiya, FA (2019). Cloud Data Provenance using IPFS and Blockchain Technology. Proceedings of the Seventh International Workshop on Security in Cloud Computing - SCC '19, 5–12. https://doi.org/10.1145/3327962.3331457

Hashem, IAT, Chang, V., Anuar, NB, Adewole, K., Yaqoob, I., Gani, A., Ahmed, E., & Chiroma, H. (2016). The role of big data in smart city. International Journal of Information Management, 36(5), 748–758. https://doi.org/10.1016/j.ijinfomgt.2016.05.002

Hennessy, G., Cook, J., Bean, M., & Dykes, K. (2011). Economic dynamics for smarter cities. 29th International Conference of the System Dynamics Society, Washington, DC. http://www.systemdynamics.org/conferences/2011/proceed/papers/P1124.pdf

Hevner, AR (2007). The Three Cycle View of Design Science Research. 19, 7.

Hussain, A., Mkpojiogu, EOC, & Jasin, N. (2017). USABILITY METRICS AND METHODS FOR PUBLIC TRANSPORTATION APPLICATIONS: A SYSTEMATIC REVIEW. 4, 9.

Hynes, N., Dao, D., Yan, D., Cheng, R., & Song, D. (2018). A demonstration of sterling: A privacy-preserving data marketplace. Proceedings of the VLDB Endowment, 11(12), 2086–2089. https://doi.org/10.14778/3229863.3236266

EIT. (2012). Intelligent Buildings: Understanding and managing the security risks. Engineering & Technology Reference. https://doi.org/10.1049/etr.2012.9001

ISO 37120. (2014). Sustainable development of communities—Indicators for city services and quality of life. International Standards Organization. https://cities.dataforcities.org/resources/ISO%2037120%20Indicators.pdf?v=1510957203519

ITU-T. (2014). Overview of key performance indicators in sustainable smart cities. International Telecommunication Union. https://www.itu.int/en/ITU-T/focusgroups/ssc/Documents/Approved_Deliverables/TS-Overview-KPI.docx

Jabbari, A., & Kaminsky, P. (2018). Blockchain and Supply Chain Management. 13.

Jang, B., Park, S., Lee, J., & Hahn, S.-G. (2018). Three Hierarchical Levels of Big-Data Market Model Over Multiple Data Sources for Internet of Things. IEEE Access, 6, 31269–31280. https://doi.org/10.1109/ACCESS.2018.2845105

Jaradat, M., Jarrah, M., Bousselham, A., Jararweh, Y., & Al-Ayyoub, M. (2015). The Internet of Energy: Smart Sensor Networks and Big Data Management for Smart Grid. Proceeded Computer Science, 56, 592-597. https://doi.org/10.1016/j.procs.2015.07.250

Jin, J., Gubbi, J., Marusic, S., & Palaniswami, M. (2014). An Information Framework for Creating a Smart City Through Internet of Things. IEEE Internet of Things Journal, 1(2), 112–121. https://doi.org/10.1109/JIOT.2013.2296516

Johnson, D., Menezes, A., & Vanstone, S. (2001). The Elliptic Curve Digital Signature Algorithm (ECDSA). International Journal of Information Security, 1(1), 36–63. https://doi.org/10.1007/s102070100002

Johnson, J., & Henderson, A. (2002). Conceptual models: Begin by designing what to design. interactions, 9(1). https://doi.org/10.1145/503355.503366

Kano, N., Seraku, N., Takahashi, F., & Tsuji, S. (1984). Attractive quality and must-be quality. Hinshitsu: The Journal of the Japanese Society for Quality Control, 14(2), 39–48.

Katuwal, GJ, Pandey, S., Hennessey, M., & Lamichhane, B. (2018). Applications of Blockchain in Healthcare: Current Landscape & Challenges. ArXiv:1812.02776 [Cs]. http://arxiv.org/abs/1812.02776

Kazi, S., Bagasrawala, M., Shaikh, F., & Sayyed, A. (2018). Smart E-Ticketing System for Public Transport Bus. 7.

Khacef, K., & Pujolle, G. (2019). Secure Peer-to-Peer Communication Based on Blockchain. In L. Barolli, M. Takizawa, F. Xhafa, & T. Enokido (Eds.), Web, Artificial Intelligence and Network Applications (Vol. 927, pp. 662–672). Springer International Publishing. https://doi.org/10.1007/978-3-030-15035-8_64

Khan, MA, & Salah, K. (2018). IoT security: Review, blockchain solutions, and open challenges. Future Generation Computer Systems, 82, 395–411. https://doi.org/10.1016/j.future.2017.11.022

Khan, Z., Anjum, A., & Kiani, SL (2013). Cloud Based Big Data Analytics for Smart Future Cities. 2013 IEEE/ACM 6th International Conference on Utility and Cloud Computing, 381–386. https://doi.org/10.1109/UCC.2013.77

Khatoun, R., & Zeadally, S. (2016). Smart cities: Concepts, architectures, research opportunities. Communications of the ACM, 59(8), 46–57.

Kim, NH, Kang, SM, & Hong, CS (2017). Mobile charger billing system using lightweight Blockchain. 2017 19th Asia-Pacific Network Operations and Management Symposium (APNOMS), 374–377. https://doi.org/10.1109/APNOMS.2017.8094151

King, S. (2013). Primecoin: Cryptocurrency with Prime Number Proof-of-Work. 6.

King, S., & Nadal, S. (2012). Ppcoin: Peer-to-peer cryptocurrency with proof-of-stake. self-published paper, August, 19.

Kitchin, R. (2014). The real-time city? Big data and smart urbanism. GeoJournal, 79(1), 1–14. https://doi.org/10.1007/s10708-013-9516-8

Koens, T., & Poll, E. (2018). The Drivers Behind Blockchain Adoption: The Rationality of Irrational Choices. 12.

Kourtit, K., Macharis, C., & Nijkamp, P. (2014). A Multi-Actor Multi-Criteria Analysis of the Performance of Global Cities. 43.

Kraft, D. (2016). Difficulty control for blockchain-based consensus systems. Peer-to-Peer Networking and Applications, 9(2), 397–413. https://doi.org/10.1007/s12083-015-0347-x

Kuechler, W., & Vaishnavi, V. (2012). A Framework for Theory Development in Design Science Research: Multiple Perspectives. 13(6), 29.

Kumar, NM, & Mallick, PK (2018). Blockchain technology for security issues and challenges in IoT. Proceeded Computer Science, 132, 1815–1823. https://doi.org/10.1016/j.procs.2018.05.140

Kwon, J. (2014). Tendermint: Consensus without Mining. 11.

Latre, S., Leroux, P., Coenen, T., Braem, B., Ballon, P., & Demeester, P. (2016). City of things: An integrated and multi-technology testbed for IoT smart city experiments. 2016 IEEE International Smart Cities Conference (ISC2), 1–8. https://doi.org/10.1109/ISC2.2016.7580875

Lazaroiu, GC, & Roscia, M. (2012). Definition methodology for the smart cities model. Energy, 47(1), 326–332. https://doi.org/10.1016/j.energy.2012.09.028

Lee, JH, Hancock, MG, & Hu, M.-C. (2014). Towards an effective framework for building smart cities: Lessons from Seoul and San Francisco. Technological Forecasting and Social Change, 89, 80–99. https://doi.org/10.1016/j.techfore.2013.08.033

Lee, JH, Phaal, R., & Lee, S.-H. (2013). An integrated service-device-technology roadmap for smart city development. Technological Forecasting and Social Change, 80(2), 286–306. https://doi.org/10.1016/j.techfore.2012.09.020

Lei A, Cruickshank H, Cao Y, Asuquo P, Ogah CPA & Sun Z (2017). Blockchain-Based Dynamic Key Management for Heterogeneous Intelligent Transportation Systems. IEEE Internet of Things Journal, 1–1. https://doi.org/10.1109/JIOT.2017.2740569

Li, F., Nucciarelli, A., Roden, S., & Graham, G. (2016). How smart cities transform operations models: A new research agenda for operations management in the digital economy. Production Planning & Control, 27(6), 514–528. https://doi.org/10.1080/09537287.2016.1147096

Li, J., Wang, J., Xu, N., Hu, Y., & Cui, C. (2018). Importance Degree Research of Safety Risk Management Processes of Urban Rail Transit Based on Text Mining Method. Information, 9(2), 26. https://doi.org/10.3390/info9020026

Li, S., Yang, L., & Gao, Z. (2015). Coordinated cruise control for high-speed train movements based on a multi-agent model. Transportation Research Part C: Emerging Technologies, 56, 281–292. https://doi.org/10.1016/j.trc.2015.04.016

Li, Y., Marier-Bienvenue, T., Perron-Brault, A., Wang, X., & Paré, G. (2018). Blockchain technology in business organizations: A scoping review. Proceedings of the 51st Hawaii International Conference on System Sciences.

Liang, G., Weller, SR, Luo, F., Zhao, J., & Dong, ZY (2018). Distributed Blockchain-Based Data Protection Framework for Modern Power Systems against Cyber Attacks. IEEE Transactions on Smart Grid, 1–1. https://doi.org/10.1109/TSG.2018.2819663

Lin, I.-C., & Liao, T.-C. (2017). A Survey of Blockchain Security Issues and Challenges. International Journal of Network Security, 19(5), 653–659. https://doi.org/10.6633/IJNS.201709.19(5).01

Linder, L., Vionnet, D., Bacher, J.-P., & Hennebert, J. (2017). Big Building Data—The Big Data Platform for Smart Buildings. Energy Procedia, 122, 589–594. https://doi.org/10.1016/j.egypro.2017.07.354

Liu, B., & Sun, X. (2018). Application Analysis of BIM Technology in Metro Rail Transit. IOP Conference Series: Earth and Environmental Science, 128, 012028. https://doi.org/10.1088/1755-1315/128/1/012028

Lu, Q.-C., & Lin, S. (2019). Vulnerability Analysis of Urban Rail Transit Network within Multi-Modal Public Transport Networks. Sustainability, 11(7), 2109. https://doi.org/10.3390/su11072109

Luo, H., Liu, C., Wu, C., & Guo, X. (2018). Urban Change Detection Based on Dempster–Shafer Theory for Multitemporal Very High-Resolution Imagery. Remote Sensing, 10(7), 980. https://doi.org/10.3390/rs10070980

Mali, N., & Kanwade, PA (2016). "A Review on Smart City through Internet of Things (IOT). International Journal of Advanced Research in Science Management and Technology, 2(6).

Mallat, N., Rossi, M., Tuunainen, VK, & Öörni, A. (2007). An empirical investigation of mobile ticketing service adoption in public transportation. Personal and Ubiquitous Computing, 12(1), 57–65. https://doi.org/10.1007/s00779-006-0126-z

Manson, N. (2006). Is operations research really research? ORiON, 22(2). https://doi.org/10.5784/22-2-40

March, ST, & Smith, GF (1995). Design and natural science research on information technology. Decision support systems, 15(4), 251–266.

Marr, B. (2018). Blockchain And The Internet Of Things: 4 Important Benefits Of Combining These Two Mega Trends. The Forbes, 2.

Mattila, J. (2016). The Blockchain Phenomenon. 27.

Mazières, D. (2015). The Stellar Consensus Protocol: 45.

Mazzanti, F., & Ferrari, A. (2018). Ten diverse formal models for a CBTC automatic train supervision system. arXiv preprint arXiv:1803.10324.

Mazzarello, M., & Ottaviani, E. (2007). A traffic management system for real-time traffic optimization in railways. Transportation Research Part B: Methodological, 41(2), 246–274. https://doi.org/10.1016/j.trb.2006.02.005

McNamee, M. (2009). Creating Smarter Cities 2011 Storylines: IBM and Smart Cities. 23.

Merkle, RC (1988). A Digital Signature Based on a Conventional Encryption Function. A Conference on the Theory and Applications of Cryptographic Techniques on Advances in Cryptology, 369–378. http://dl.acm.org/citation.cfm?id=646752.704751

Mohamad Noor, M. binti, & Hassan, WH (2019). Current research on Internet of Things (IoT) security: A survey. Computer Networks, 148, 283–294. https://doi.org/10.1016/j.comnet.2018.11.025

Mohammadi, M., Al-Fuqaha, A., Guizani, M., & Oh, J.-S. (2018). Semisupervised Deep Reinforcement Learning in Support of IoT and Smart City Services. IEEE Internet of Things Journal, 5(2), 624–635. https://doi.org/10.1109/JIOT.2017.2712560

Morandi, C., Rolando, A., & Di Vita, S. (2016). From Smart City to Smart Region. Springer International Publishing. https://doi.org/10.1007/978-3-319-17338-2

Moreno, MV, Terroso-Saenz, F., Gonzalez-Vidal, A., Valdes-Vela, M., Skarmeta, AF, Zamora, MA, & Chang, V. (2017). Applicability of Big Data Techniques to Smart Cities Deployments. IEEE Transactions on Industrial Informatics, 13(2), 800–809. https://doi.org/10.1109/TII.2016.2605581

Moreno Pires, S., Fidélis, T., & Ramos, TB (2014). Measuring and comparing local sustainable development through common indicators: Constraints and achievements in practice. Cities, 39, 1–9. https://doi.org/10.1016/j.cities.2014.02.003

Nair, V., Pawar, A., Tidke, DL, Pagar, V., & Wani, N. (2018). Online Bus Tracking and Ticketing System. 4.

Nakamoto, S. (2008). Bitcoin: A Peer-to-Peer Electronic Cash System. 9.

Nam, T., & Pardo, TA (2011). Conceptualizing smart city with dimensions of technology, people, and institutions. Proceedings of the 12th annual international digital government research conference: digital government innovation in challenging times, 282–291. http://dl.acm.org/citation.cfm?id=2037602

Neirotti, P., De Marco, A., Cagliano, AC, Mangano, G., & Scorrano, F. (2014). Current trends in Smart City initiatives: Some styled facts. Cities, 38, 25–36. https://doi.org/10.1016/j.cities.2013.12.010

Ng, ST, Xu, FJ, Yang, Y., & Lu, M. (2017). A Master Data Management Solution to Unlock the Value of Big Infrastructure Data for Smart, Sustainable and Resilient City Planning. Procedia Engineering, 196, 939-947. https://doi.org/10.1016/j.proeng.2017.08.034

Niemeyer, M., Henneböhle, K., & Kuller, M. (2014). Security requirements of IoT-based smart buildings using RESTful Web Services. Acta Polytechnica Hungarianica, 10.

Nitti, M., Pilloni, V., Giusto, D., & Popescu, V. (2017). IoT Architecture for a Sustainable Tourism Application in a Smart City Environment. Mobile Information Systems, 2017, 1–9. https://doi.org/10.1155/2017/9201640

Otero-Cerdeira, L., Rodríguez-Martínez, FJ, & Gómez-Rodríguez, A. (2014). Enhancing Alignment Results in Ontology Matching for Smart Cities. WOIS@BIR, 55–65. ftp://ceur-ws.org/pub/publications/CEUR-WS/Vol-1230.zip#page=60

Peffers, K., Tuunanen, T., Rothenberger, MA, & Chatterjee, S. (2007). The Design Science Research Methodology for Information Systems Research. Journal of Management Information Systems, 24(3), 45–77. https://doi.org/10.2753/MIS0742-1222240302

Pellegrino, JL, Fanney, AH, Bushby, ST, Domanski, PA, Healy, WM, & Persily, AK (2010). Measurement Science Roadmap for Net-Zero Energy Buildings (NIST Technical Note 1660; Workshop Summary Report). National Institute of Standards and Technology. https://ws680.nist.gov/publication/get_pdf.cfm?pub_id=905024

Petrolo, R., Loscrì, V., & Mitton, N. (2017). Towards a smart city based on cloud of things, a survey on the smart city vision and paradigms: R. Petrolo, V. Loscrì and N. Mitton. Transactions on Emerging Telecommunications Technologies, 28(1), e2931. https://doi.org/10.1002/ett.2931

Petrolo, R., Loscrì, V., & Mitton, N. (2012). Towards a Smart City based on Cloud of Things, a survey on the smart city vision and paradigms. 12.

Pillmann, J., Wietfeld, C., Zarcula, A., Raugust, T., & Alonso, DC (2017). Novel common vehicle information model (CVIM) for future automotive vehicle big data marketplaces. 2017 IEEE Intelligent Vehicles Symposium (IV), 1910–1915. https://doi.org/10.1109/IVS.2017.7995984

Popescul, D., & Radu, LD (2016). Data Security in Smart Cities: Challenges and Solutions. Informatica Economica, 20(1/2016), 29–38. https://doi.org/10.12948/issn14531305/20.1.2016.03

Prehofer, C., van Gurp, J., Stirbu, V., Satish, S., Tarkoma, S., di Flora, C., & Liimatainen, PP (2010). Practical Web-Based Smart Spaces. IEEE Pervasive Computing, 9(3), 72–80. https://doi.org/10.1109/MPRV.2009.88

Profanter, S., Tekat, A., Dorofeev, K., & Rickert, M. (2019). OPC UA versus ROS, DDS, and MQTT: Performance Evaluation of Industry 4.0 Protocols.

Proceedings of the IEEE International Conference on Industrial Technology (ICIT)., 8.

Psomakelis, E., Aisopos, F., Litke, A., Tserpes, K., Kardara, M., & Campo, PM (2016). Big IoT and Social Networking Data for Smart Cities—Algorithmic Improvements on Big Data Analysis in the Context of RADICAL City Applications: Proceedings of the 6th International Conference on Cloud Computing and Services Science, 396–405. https://doi.org/10.5220/0005934503960405

Purao, S. (2002). Design Research in the Technology of Information Systems: Truth or Dare. 37.

Qu, C., Tao, M., Zhang, J., Hong, X., & Yuan, R. (2018). Blockchain Based Credibility Verification Method for IoT Entities. Security and Communication Networks, 2018, 1–11. https://doi.org/10.1155/2018/7817614

Ra, G.-J., & Lee, I.-Y. (2018). A Study on KSI-based Authentication Management and Communication for Secure Smart Home Environments. KSII Transactions on Internet and Information Systems, 12(2). https://doi.org/10.3837/tiis.2018.02.021

Rabah, K. (2018). Convergence of AI, IoT, Big Data and Blockchain: A Review. 1(1), 18.

Rao, JS, & Syamala, M. (2017). Internet of Things (IoT) based Smart City Architecture and its Applications. Mathematical Sciences, 6(10), 6.

Rathore, MM, Ahmad, A., Paul, A., & Rho, S. (2016). Urban planning and building smart cities based on the Internet of Things using Big Data analytics. Computer Networks, 101, 63–80. https://doi.org/10.1016/j.comnet.2015.12.023

Rathore, MM, Paul, A., Ahmad, A., & Jeon, G. (2017). IoT-Based Big Data: From Smart City towards Next Generation Super City Planning. International Journal on Semantic Web and Information Systems, 13(1), 28–47. https://doi.org/10.4018/IJSWIS.2017010103

Rey-Robert, X. (2009). Smarter Cities – Dublin event. 56.

Rhee, S. (2016). Catalyzing the internet of things and smart cities: Global city teams challenge. Science of Smart City Operations and Platforms Engineering (SCOPE) in partnership with Global City Teams Challenge (GCTC)(SCOPE-GCTC), 2016 1st International Workshop on, 1–4.

Rinaldi, S., Ferrari, P., & Flammini, A. (2017). Analysis of modular bridge platform for heterogeneous software defined networking in smart city applications. 2017 IEEE International Instrumentation and Measurement Technology Conference (I2MTC), 1–6. https://doi.org/10.1109/I2MTC.2017.7969846

Risius, M., & Spohrer, K. (2017). A Blockchain Research Framework: What We (don't) Know, Where We Go from Here, and How We Will Get There. Business & Information Systems Engineering, 59(6), 385–409. https://doi.org/10.1007/s12599-017-0506-0

Rithika, G., Harshitha, PS, Manasa, V., & Anisha, MPR (2019). Blockchain-Foundational Technology to Change the World. 1076, 12.

Rivest, RL, Shamir, A., & Adleman, L. (1978). A method for obtaining digital signatures and public-key cryptosystems. Communications of the ACM, 21(2), 120–126.

Roberts, S., Bonenberg, L., Meng, X., Moore, T., & Hill, C. (2017). Predictive Intelligence for a Rail Traffic Management System. 9.

Roche, S., Nabian, N., Kloeckl, K., & Ratti, C. (2012). Are 'Smart Cities' Smart Enough? 22.

Roman-Castro, R., Lopez, J., & Gritzalis, S. (2018). Evolution and Trends in IoT Security. Computer, 51(7), 16–25. https://doi.org/10.1109/MC.2018.3011051

Ruta, M., Scioscia, F., Ieva, S., Capurso, G., Loseto, G., Gramegna, F., Pinto, A., & Sciascio, ED (2017). Semantic-enhanced blockchain technology for smart cities and communities. two.

Sadoghi, M. (2017). ExpoDB: An Exploratory Data Science Platform. 1.

Samuel, SSI (2016). A review of connectivity challenges in IoT-smart home. Big Data and Smart City (ICBDSC), 2016 3rd MEC International Conference on, 1–4.

Sánchez, L., Elicegui, I., Cuesta, J., Muñoz, L., & Lanza, J. (2013). Integration of Utilities Infrastructures in a Future Internet Enabled Smart City Framework. Sensors, 13(11), 14438–14465. https://doi.org/10.3390/s131114438

Sanchez, L., Galache, JA, Gutierrez, V., Manuel, J., Bernat, J., Gluhak, A., & Garcia, T. (2011). SmartSantander: The meeting point between Future Internet research and experimentation and the smart cities. 9.

Santana, EFZ, Chaves, AP, Gerosa, MA, Kon, F., & Milojicic, D. (2016). Software Platforms for Smart Cities: Concepts, Requirements, Challenges, and a Unified Reference Architecture. ArXiv:1609.08089 [Cs]. http://arxiv.org/abs/1609.08089

Santos, AP (2015). Cities of the Future: Talent, Innovation and Collaboration. 20.

Schaffers, H., Komninos, N., Pallot, M., Trousse, B., Nilsson, M., & Oliveira, A. (2011). Smart Cities and the Future Internet: Towards Cooperation Frameworks for Open Innovation. In J. Domingue, A. Galis, A. Gavras, T. Zahariadis, D. Lambert, F. Cleary, P. Daras, S. Krco, H. Müller, M.-S. Li, H. Schaffers, V. Lotz, F. Alvarez, B. Stiller, S. Karnouskos, S. Avessta, & M. Nilsson (Eds.), The Future Internet (Vol. 6656, pp. 431-446). Springer Berlin Heidelberg. https://doi.org/10.1007/978-3-642-20898-0_31

Schelini, PW (2006). Theory of fluid and crystallized intelligences: Inception and evolution. Psychology Studies (Natal), 11(3), 323–332. https://doi.org/10.1590/S1413-294X2006000300010

Schleicher, JM, Vogler, M., Dustdar, S., & Inzinger, C. (2016). Enabling a Smart City Application Ecosystem: Requirements and Architectural Aspects. IEEE Internet Computing, 20(2), 58–65. https://doi.org/10.1109/MIC.2016.39

Schwartz, D., Youngs, N., & Britto, A. (2014). The Ripple Protocol Consensus Algorithm. 8.

Scuotto, V., Caputo, F., Villasalero, M., & Del Giudice, M. (2017). A multiple buyer – supplier relationship in the context of SMEs' digital supply chain management.

Production Planning & Control, 28(16), 1378–1388. https://doi.org/10.1080/09537287.2017.1375149

Sewell, JE, & Fraser, DJ (2019). A Conceptual and literature review of the effectiveness BREEAM. 12.

Shaghaghi, A., Kaafar, MA, Buyya, R., & Jha, S. (2018). Software-Defined Network (SDN) Data Plane Security: Issues, Solutions and Future Directions. ArXiv:1804.00262 [Cs]. http://arxiv.org/abs/1804.00262

Shaikh, T., Ismail, S., & Stevens, JD (2016). Aura Minora: A user centric IOT architecture for Smart City. Proceedings of the International Conference on Big Data and Advanced Wireless Technologies - BDAW '16, 1–5. https://doi.org/10.1145/3010089.3016028

Sharma, A., & Bhuriya, D. (2019). Literature Review of Blockchain Technology. 6(1), 8.

Sheikh, N., Khapekar, Ku. T., Kumar, S., & Kumar, V. (2018). Techniques of E-ticket System: A Review. 3(1), 4.

Shyam R., Ganesh HB, B., Kumar S., S., Poornachandran, P., & Soman KP (2015). Apache Spark a Big Data Analytics Platform for Smart Grid. Procedia Technology, 21, 171–178. https://doi.org/10.1016/j.protcy.2015.10.085

Singhal, S., McGreal, S., & Berry, J. (2013). Application of a hierarchical model for city competitiveness in cities of India. Cities, 31, 114–122.

Smith, G., Ofe, HA, & Sandberg, J. (2016). Digital Service Innovation from Open Data: Exploring the Value Proposition of an Open Data Marketplace. 2016 49th Hawaii International Conference on System Sciences (HICSS), 1277–1286. https://doi.org/10.1109/HICSS.2016.162

Söderström, O., Paasche, T., & Klauser, F. (2014). Smart cities as corporate storytelling. City, 18(3), 307–320. https://doi.org/10.1080/13604813.2014.906716

Sompolinsky, Y., & Zohar, A. (2013). Accelerating Bitcoin's Transaction Processing. 31.

Sonawane, SA, & Shaikh, SA (2017). Implementing Smart City Concept with Various Application Using IOT Based Technique.

Song, L., Li, Q., List, G., Deng, Y., & Lu, P. (2017). Using an AHP-ISM Based Method to Study the Vulnerability Factors of Urban Rail Transit System. Sustainability, 9(6), 1065. https://doi.org/10.3390/su9061065

Sousa, J., Bessani, A., & Vukolić, M. (2017). A Byzantine Fault-Tolerant Ordering Service for the Hyperledger Fabric Blockchain Platform. arXiv preprint arXiv:1709.06921.

Spiekermann, S., & Cranor, LF (2009). Engineering Privacy. IEEE Transactions on Software Engineering, 35(1), 67–82. https://doi.org/10.1109/TSE.2008.88

Stanford-Clark, A., & Truong, HL (1999). MQTT For Sensor Networks (MQTT-SN) Protocol Specification. 28.

Stephen, R., & Alex, A. (2018). A Review on BlockChain Security. IOP Conference Series: Materials Science and Engineering, 396, 012030. https://doi.org/10.1088/1757-899X/396/1/012030

Strass, G., & Williamson, J. (2014). Five Best Practices to Improve Building Management Systems (BMS) Cyber Security. Schneider Electric White Paper, 11.

Streitz, NA, Rocker, C., Prante, T., van Alphen, D., Stenzel, R., & Magerkurth, C. (2005). Designing smart artifacts for smart environments. Computer, 38(3), 41–49. https://doi.org/10.1109/MC.2005.92

Suarez-Albela, M., Fernandez-Carames, TM, Fraga-Lamas, P., & Castedo, L. (2018). A Practical Performance Comparison of ECC and RSA for Resource-Constrained IoT Devices. 2018 Global Internet of Things Summit (GIoTS), 1–6. https://doi.org/10.1109/GIOTS.2018.8534575

Teixeira, J., Fernandes, P., Bandeira, JM, & Coelho, MC (2017). Information Management for Smart and Sustainable Mobility. 19.

Telles, MJ (2017). A Computational Model for Assistive Smart Cities. 10(1), 28.

Temple, WG, Li, Y., Tran, BAN, Liu, Y., & Chen, B. (2017). Railway System Failure Scenario Analysis. In G. Havarneanu, R. Setola, H. Nassopoulos, & S. Wolthusen (Eds.), Critical Information Infrastructures Security (Vol. 10242, pp. 213–225). Springer International Publishing. https://doi.org/10.1007/978-3-319-71368-7_18

Travizano, M., Minnoni, M., Ajzenman, G., Sarraute, C., & Penna, ND (2018). Wibson: A decentralized marketplace empowering individuals to safely monetize their personal data. 18.

Trindade, EP, Hinnig, MPF, da Costa, EM, Marques, JS, Bastos, RC, & Yigitcanlar, T. (2017). Sustainable development of smart cities: A systematic review of the literature. Journal of Open Innovation: Technology, Market, and Complexity, 3(1). https://doi.org/10.1186/s40852-017-0063-2

Tyrinopoulos, Y., & Aifadopoulou, G. (2008). A complete methodology for the quality control of passenger services in the public transport business. 38, 16.

Uceda-Sosa, R., Srivastava, B., & Schloss, RJ (2011). Building a highly consumable semantic model for smarter cities. Proceedings of the AI for an Intelligent Planet on - AIIP '11, 1–8. https://doi.org/10.1145/2018316.2018319

IGU. (2004). Urban Governance Index (UGI): Methodology Guidelines. UN-Habitat. http://mirror.unhabitat.org/downloads/docs/2232_55927_Addendum%20-%20Methodology%20Guidelines.doc

UN-Habitat. (2013). STATE OF THE WORLD'S CITIES 2012/2013—Prosperity of cities. Routledge [ua]. http://mirror.unhabitat.org/pmss/getElectronicVersion.aspx?nr=3387&alt=1

Vaishnavi, VK, Vaishnavi, VK, & Kuechler, W. (2015). Design Science Research Methods and Patterns: Innovating Information and Communication Technology, 2nd Edition (0 ed.). CRC Press. https://doi.org/10.1201/b18448

van Lierop, D., & El-Geneidy, A. (2016). Enjoying loyalty: The relationship between service quality, customer satisfaction, and behavioral intentions in public transit.

Research in Transportation Economics, 59, 50–59. https://doi.org/10.1016/j.retrec.2016.04.001

Vasin, P. (2014). BlackCoin's Proof-of-Stake Protocol v2. two.

Venable, JR (2006). The Role of Theory and Theorising in Design Science Research. 18.

Venable, JR, Pries-Heje, J., & Baskerville, R. (2017). Choosing a Design Science Research Methodology. 11.

Vilajosana, I., Llosa, J., Martinez, B., Domingo-Prieto, M., Angles, A., & Vilajosana, X. (2013). Bootstrapping smart cities through a self-sustainable model based on big data flows. IEEE Communications Magazine, 51(6), 128–134.

Vujicic, D., Jagodic, D., & Randic, S. (2018). Blockchain technology, bitcoin, and Ethereum: A brief overview. 2018 17th International Symposium INFOTEH-JAHORINA (INFOTEH), 1–6. https://doi.org/10.1109/INFOTEH.2018.8345547

Walls, JG, Widmeyer, GR, & El Sawy, OA (1992). Building an information system design theory for vigilant EIS. Information systems research, 3(1), 36–59.

Walters, D. (2011). Smart cities, smart places, smart democracy: Form-based codes, electronic governance and the role of place in making smart cities. Intelligent Buildings International, 3(3), 198–218. https://doi.org/10.1080/17508975.2011.586670

Wang, Haifeng, Zhao, N., Ning, B., Tang, T., & Chai, M. (2018). Safety monitor for train-centric CBTC system. IET Intelligent Transport Systems, 12(8), 931–938.

Wang, Huaiqing, Chen, K., & Xu, D. (2016). The maturity model for blockchain adoption. 5.

Wang, W., Hoang, DT, Xiong, Z., Niyato, D., Wang, P., Hu, P., & Wen, Y. (2018). A Survey on Consensus Mechanisms and Mining Management in Blockchain Networks. ArXiv:1805.02707 [Cs]. http://arxiv.org/abs/1805.02707

Wei, Y., Huang, C., Li, J., & Xie, L. (2016). An evaluation model for urban carrying capacity: A case study of China's mega-cities. Habitat International, 53, 87–96. https://doi.org/10.1016/j.habitatint.2015.10.025

Wohrer, M., & Zdun, U. (2018). Smart contracts: Security patterns in the ethereum ecosystem and solidity. 2018 International Workshop on Blockchain Oriented Software Engineering (IWBOSE), 2–8. https://doi.org/10.1109/IWBOSE.2018.8327565

Wolisz, H., Böse, L., Harb, H., Streblow, R., & Müller, D. (2014). CITY DISTRICT INFORMATION MODELING AS A FOUNDATION FOR SIMULATION AND EVALUATION OF SMART CITY APPROACHES. 9.

Wood, G. (2019). ETHEREUM: A SECURE DECENTRALIZED GENERALIZED TRANSACTION LEDGER. 39.

Wust, K., & Gervais, A. (2018). Do you Need a Blockchain? 2018 Crypto Valley Conference on Blockchain Technology (CVCBT), 45–54. https://doi.org/10.1109/CVCBT.2018.00011

Xu, X., Weber, I., Staples, M., Zhu, L., Bosch, J., Bass, L., Pautasso, C., & Rimba, P. (2017). A Taxonomy of Blockchain-Based Systems for Architecture Design.

2017 IEEE International Conference on Software Architecture (ICSA), 243–252. https://doi.org/10.1109/ICSA.2017.33

Yan, F., Gao, C., Tang, T., & Zhou, Y. (2017). A Safety Management and Signaling System Integration Method for Communication-Based Train Control System. Urban Rail Transit, 3(2), 90–99. https://doi.org/10.1007/s40864-017-0051-7

Yang, Y., Wu, L., Yin, G., Li, L., & Zhao, H. (2017). A Survey on Security and Privacy Issues in Internet-of-Things. IEEE Internet of Things Journal, 4(5), 1250–1258. https://doi.org/10.1109/JIOT.2017.2694844

Yli-Huumo, J., Ko, D., Choi, S., Park, S., & Smolander, K. (2016). Where Is Current Research on Blockchain Technology?—The Systematic Review. PloS one, 11(10), e0163477.

Yuan, Y., & Wang, F.-Y. (2016). Blockchain-based intelligent transportation systems towards. Intelligent Transportation Systems (ITSC), 2016 IEEE 19th International Conference on, 2663–2668.

Zamfir, V. (2015). Introducing Casper "the Friendly Ghost". https://blog.ethereum.org/2015/08/01/introducing-casper-friendly-ghost/

Zhang K, Ni J, Yang K, Liang X, Ren J & Shen XS (2017). Security and Privacy in Smart City Applications: Challenges and Solutions. IEEE Communications Magazine, 55(1), 122–129. https://doi.org/10.1109/MCOM.2017.1600267CM

Zhang, M. (2017). DECISION SUPPORT APPROACH FOR INTEGRATED MAINTENANCE PROGRAM OF URBAN RAIL TRANSIT. International Journal of Computing, 16(3), 143–151.

Zheng, Q., Li, Y., Chen, P., & Dong, X. (2018). An Innovative IPFS-Based Storage Model for Blockchain. 2018 IEEE/WIC/ACM International Conference on Web Intelligence (WI), 704–708. https://doi.org/10.1109/WI.2018.000-8

Zheng, Z., Xie, S., Dai, H., Chen, X., & Wang, H. (2017a). An Overview of Blockchain Technology: Architecture, Consensus, and Future Trends. 2017 IEEE International Congress on Big Data (BigData Congress), 557–564. https://doi.org/10.1109/BigDataCongress.2017.85

Zheng, Z., Xie, S., Dai, H., Chen, X., & Wang, H. (2017b). An Overview of Blockchain Technology: Architecture, Consensus, and Future Trends. 2017 IEEE International Congress on Big Data (BigData Congress), 557–564. https://doi.org/10.1109/BigDataCongress.2017.85

Zhou, K., Fu, C., & Yang, S. (2016). Big data driven smart energy management: From big data to big insights. Renewable and Sustainable Energy Reviews, 56, 215–225. https://doi.org/10.1016/j.rser.2015.11.050

Zygiaris, S. (2013). Smart City Reference Model: Assisting Planners to Conceptualize the Building of Smart City Innovation Ecosystems. Journal of the Knowledge Economy, 4(2), 217–231. https://doi.org/10.1007/s13132-012-0089-4

Zyskind, G., Nathan, O., & Pentland, A. «Sandy». (2015). Decentralizing Privacy: Using Blockchain to Protect Personal Data. 180–184. https://doi.org/10.1109/SPW.2015.27

ANNEX I

Annex I - Comparative Table of Blockchain Platforms

Name	Link	Consensus	main functions
AWS Blockchain	https://aws.amazon.com/en/blockchain/		It offers a solution through Blockchain platforms: Ethereum, Hyperledger Fabric, Quorum and Corda. AWS Blockchain Templates for Ethereum and for Hyperledger Fabric pay for use Security for businesses, with users being able to add permissions to control access to AWS resources. Access resource activity through AWS CloudTrail. With AWS blockchain Templates, users can quickly implement secure blockchain networks.
Azure BaaS (Blockchain as a Service)	https://azure.microsoft.com/en-in/solutions/blockchain/		It offers a solution through Blockchain platforms: Ethereum, Hyperledger Fabric, Quorum, Chain and Corda. Azure blockchain template for Ethereum and for Hyperledger Fabric Pay-as-you-go. It offers pre-configured networks and infrastructure that allow users to immediately start developing decentralized applications. Through internal links to Azure tools, they can validate and interact with blockchain developments faster and easier. Cloud platform solutions are scalable.
BigChainDB	https://www.bigchaindb.com/	Federation of nodes with voting permissions	Each record is recorded in the blockchain database without the need for Merkle trees. Support for custom assets, transactions, permissions and transparency. Consensus Model Federation (federation of nodes vote). Supports public and private networks. Has no native currency - any asset, symbol or currency can be issued. Set transaction-level permissions. It's open source.
BitShares	https://bitshares.org/	Delegated Proof-of-Stake (dPoS) Consensus	The platform boasts of supersonic transaction speed, low cost and top-notch security. A user-friendly interface, code-friendly and extensible, with the characteristics of financial smart contracts, and a robust testing infrastructure designed to avoid bugs.
chain	https://chain.com/	Federated Consensus	Blockchain platform that suitable for financial applications. There is an open-source development edition. Network type with permission Licensing with prices for companies Supported Language Java, Ruby, Node.js Medium Popularity but actively updated on GitHub GitHub Repo SDK-Java (Java) sdk-Ruby (Ruby) sdk-NodeJS (Node.js / Javascript)
Chain Core	http://core-chain.io/	Federated Consensus	Native digital assets - currencies, bonds, etc. Role-based permissions for operating, accessing, and participating in a network. Support for multi-signature accounts. Federated Consensus. Support for smart contracts. transaction privacy
CoinList	https://coinlist.co/	Uses unnecessary proof of work (PoW) consensus.	The objective is to clearly define the nature of an Initial Public Offering (ICO- *Initial Coin Offering*), therefore, minimize the possibility of violating regulations, such as those established by the Securities and Exchange Commission (SEC). Filecoin is like bitcoin, but with storage instead of hashers. The new Proof of Replication function creates a useful and valuable storage service as a by-product of the mining process.
Rope	https://www.corda.net	Transactions confirmed individually by each of the participants in a transaction.	No global data transmission over the network. Pluggable consensus. Query with SQL, allowing to join external databases, bulk imports.
credits	https://credits.works/	Credits uses a Proof of Stake (PoS) variant	*framework* of development for creating distributed books with permission. Consensus is based on an algorithm with variable voting power.
Domus Tower Blockchain	http://domustower.com/	Any agent that has access to a blockchain has 100% authority to write transactions to that chain.	Creating linked blockchains in which the assets on one account on one blockchain must match the liabilities on the account on another blockchain. Ability to record at a high rate of transactions in a scalable way. Double-entry balance recording tracks credits and debits.

Name	Link	Consensus	main functions
		Authority is centralized under this model.	
Blockchain Platform Elements	https://elementsproject.org/	Asset labels are not part of the network protocol consensus and are local only to each node. The hexadecimal value of the asset is used, which is shared across the network.	Confidential Assets - Issues multiple assets with identifiers and values that are "blind" and auditable. Confidential Transactions - keep transferred amounts visible only to transaction participants and designated entities. Additional opcodes - include preconditioned opcodes (including concatenation and subsequences, integer changes, and various bitwise operations), operation that produces a random number within a range from a seed, and operation that checks a signature against a message on the stack, rather than the spend operation itself. Cross-blockchain operations to be built in a decentralized way with tokens to be moved from one blockchain to another. Signed Blocks - Allows blocks to be cryptographically signed so you can verify your identity in the future Witness segregation. Bitcoin transactions contain information about the effect on accounting and data that proves that the transaction is authorized. Using witness segregation, transaction IDs are redefined to rely only on effect information and blocks and to commit separately to witness data, avoiding transaction manipulation. Relative Time Lock that allows a transaction to be locked in time.
EOS	https://eos.io/	DPoS (Delegated Proof-of-Stake) and aBFT (Byzantine Fault Tolerance) consensus	Platform designed to allow large corporations, banks and governments to build decentralized applications.
Eris	https://erisindustries.com/	Tendermint consensus	Depends on the types you want to turn on and off. In addition, it will help to directly develop a blockchain group with permission for certain individuals. In addition to smart contract tools, it allows you to analyze action step by step. This software allows anyone present on the platform to create and run an application from anywhere. Any entity can use smart contracts to do business with Eris automatically. It's open source.
Ethereum	https://www.ethereum.org/	Initially PoW and evolved PoS in version 2.0	Ethereum is an open blockchain platform that allows anyone to create and use decentralized applications that run on blockchain technology. Ethereum is adaptable and flexible. Popularity and activity is high and actively followed on GitHub The network type is public, it uses smart contracts based on the Ether currency. Supported Languages - Python, Go, C++ GitHub repo - pyethereum (Python) gpethereum (golang) CPP-ethereum (C++) It provides a decentralized virtual machine called the Ethereum Virtual Machine (EVM) to complete scripts using a global network of common nodes.
Hydrachain	https://github.com/HydraChain/hydrachain	Byzantine fault tolerant consensus protocol	HydraChain is joint development of BrainBot technologies and the Ethereum project. HydraChain is an extension of the Ethereum platform that supports the creation of scalable blockchain-based applications that meet organizational and regulatory requirements. Popularity - low but active on GitHub Network type - private / permissioned Open Source Supported Languages - Python, GitHub Repo - hydrachain (Python)
Hyperledger Fabric	https://github.com/hyperledger/fabric	Consensus based on PoS (Proof of Stake)	Journal (legder) query and update using key-based, range-based, and composite-key searches. History queries only. The operations contain the signatures of the peers to be endorsed. A channel's accounting contains a configuration block, policy definition, access control lists, and other pertinent information. The channel allows encryption to come from different certificate authorities
Hyperledger Iroha	https://www.hyperledger.org/projects/iroha	*Sumeragi*, Byzantine fault-tolerant consensus algorithm heavily inspired by the B-	"Simple and modular" the distributed ledger system and prominently in the development of mobile applications.

Name	Link	Consensus	main functions
		blockchain algorithm	
Hyperledger Sawtooth Lake	https://sawtooth.hyperledger.org/	Elapsed Time Test	Hyperledger is an open source collaborative project created to advance blockchain technologies, across multiple industries: finance, banking, IoT, supply chains, manufacturing and technology. Popularity - high and actively updated on GitHub Network type - both private and public *Open Source* Supported Languages - Python (for Sawtooth Lake)
IBM Bluemix Blockchain	https://console.ng.bluemix.net/catalog/services/blockchain/	raft consensus	IBM has also released its blockchain platform which is available as part of the Bluemix service catalog. It uses the HyperLedger design and provides additional security and infrastructure facilities for businesses. Popularity - Medium but actively updated on GitHub Network Type - Private / Permissioned Pricing - limited free plan and paid upgrade to enterprise plan, Supported languages - GO, Javascript, ibm-blockchain-js (Javascript)
*IOT*THE	https://www.IoTa.org http://IoTatoken.com/	IOTA consensus from the same family of consensus algorithms as Snowball	*IOTA* deviates from type blockchain projects. Highly scalable, where increased network activity decreases transaction settlement times. Low requirements on resources, designed for small devices such as sensors to participate. Zero fee transactions. Secure data transfer, where data is encrypted, enabling secure data transfer, storage and reference. Offline transactions, where devices do not need always-on connectivity Immune to quantum technology, as it uses special signatures, which makes it resilient to the next generation of quantum computing. Low Popularity but actively updated on GitHub Network type - public, with permission. Supported Languages Python, C, Javascript GitHub Repo IoTa.lib.py (Java) ccurl (C) IoTa.lib.js (Javascript)
KICKICO	https://www.kickico.com/	Management will be carried out by voting, with a 51% consensus among voters.	Blockchain platform built using Ethereum-based smart contracts. It offers comprehensive solutions for ICOs, crowdfunding, and crowdinvesting using blockchain technology.
Monax/Hyperledger Burrow	https://monax.io/	*Proof of Stake* (Tendermint)	It bases its operation on three functions: create, track and prove. To create it uses a contract template. To track it uses digital workflows and integrations, tracks deliveries and activates automated actions. To prove it, it uses sophisticated reports to prove compliance with contractual requirements and contractual activity, executes the digitized contract and feeds its business with embedded contracts, such as: non-disclosure agreements (NDA-Non-Disclosure Agreements), service contracts (MSA-Master Services Agreements), declarations of work (SOW), change orders (CO-*change orders*) and Terms of Service (ToS-Terms of Service).
multichain	https://www.multicchain.com/	Distributed consensus among the validators of the identified block. Identical to practical Byzantine fault tolerance, with one validator per block.	Native multi-currency support. Permission Management. Quick deployment. Multiple networks can be simultaneously on a single server. Custom parameter per network (allowed transaction types, confirmation times, minimum quantities, transaction fee and size limits). This software helps design, implement and operate distributed journals. This platform allows you to create and deploy private blockchains (with blockchains permission) within or between organizations. Provides privacy and control within a peer-to-peer private network. It is an improved version of the bitcoin core software for private financial transactions. Popularity - Medium but active on GitHub Network type - Private, with permission *Open Source* Supported Languages - Python, C#, JavaScript, PHP, Ruby, GitHub Repo - savior (Python); C# MultichainLib (C#); Multichain-Node (JavaScript); libphp-multichain (PHP);

Name	Link	Consensus	main functions
			multichain-client (Ruby); savior (Python)
NOR	https://nem.io/	*proof of importance* (POI)	The purpose of the exchange is to provide a dedicated platform for NEM-based ICOs and token trading. Investors can buy and sell XEM, the NEM cryptocurrency and have the option to exchange.
NXT	https://nxtplatform.org/	Voting for Nxt is important to reach consensus.	The decentralized exchange of assets, in decentralized voting and governance systems, a manageable blockchain and phased operations.
Open Chain	https://www.openchain.org/	partitioned consensus	Suitable for organizations that issue and manage digital assets. It follows a partitioned consensus system, where each Openchain instance only has one authority to validate transactions, depending on the resources being exchanged. The hierarchical account system allows you to set permissions at any level. A client-server (centralized) architecture can be more efficient and reliable than a peer-to-peer architecture. Popularity - Average. Active on GitHub Network Type - Private *Open Source* Supported Languages - JavaScript GitHub Repo - openchain-js (Javascript)
Oracle	https://cloud.oracle.com/en_US/blockchain	Envisioned on the Hyperledger Fabric platform	Pay per use. It offers an autonomous, self-safe and self-repairing solution. Allows you to develop the blockchain network and applications on the Hyperledger Fabric framework. Through Oracle Integration Cloud you can integrate local applications, Oracle SaaS and third-party applications. Enterprise Grade blockchain Solution provides enterprise-grade features such as built-in backup, enhanced security, role-based identity management, and certificate revocation management. Oracle blockchain is easy to implement as it offers a fully manageable and pre-built blockchain service.
Quorum	https://www.goquorum.com/	Algorithms:RAFT, such as the Proof of Stake; IBFT (Istanbul Byzantine Fault Tolerant); or PoA (Proof of Authority)	It features strong privacy in that private information is never transmitted to network participants. Private data is encrypted and shared between interested parties. Consensus: RAFT, IBFT (Istanbul Byzantine Fault Tolerant) or. PoA (Proof of Authority) consensus model with identity in participation. Permission Flexibly define how to incorporate entities and users into your network using Quorum's "smart" contract-based permissions model. Map sub-entities to parent entities to suit your organization's structure. simple integration With quorum-cloud, you can easily deploy Quorum across different cloud environments. Or use Docker for seamless integration between environments. Tool Compatibility: TruffleSuite[71], MetaMask[72], Remix[73] and OpenZeppelin[74]. *Open Source*
stellar	https://www.stellar.org	Stellar Consensus Protocol	Stellar is a distributed payments infrastructure that connects banks, payment systems and people. Stellar allows the construction of mobile wallets, banking tools and "smart" devices. It provides HTTP API RESTful servers called Horizon, which connect to Stellar Core, the foundation of the Stellar network. *Open Source*
Symbiont Assembly	https://symbiont.io/technology/introducing-symbiont-assembly/	Byzantine fault tolerance	Ability to handle thousands of transactions per second. Assembly API - Standard JSON over HTTP.

---

[71] Truffle - Tools for smart contracts - https://www.trufflesuite.com/ , accessed on 15-04-2017
[72] Metamask - Embeds Ethereum in your browser - https://metamask.io/ , accessed on 15-04-2017
[73] Remix - Ethereum IDE - https://remix.ethereum.org/ , accessed on 15-04-2017
[74] OpenZeppelin - Framework for creating secure smart contracts - https://openzeppelin.org/, accessed on 15-04-2017